THE
HISTORY
OF
ISSUES

Evolution

THE
HISTORY
OF
ISSUES

Evolution

Don Nardo, *Book Editor*

Bruce Glassman, *Vice President*
Bonnie Szumski, *Publisher*
Helen Cothran, *Managing Editor*

GREENHAVEN PRESS
An imprint of Thomson Gale, a part of The Thomson Corporation

THOMSON

GALE

Detroit • New York • San Francisco • San Diego • New Haven, Conn.
Waterville, Maine • London • Munich

3 1257 01625 0440

THOMSON
——— ✳ ———™
GALE

© 2005 Thomson Gale, a part of The Thomson Corporation.

Thomson and Star Logo are trademarks and Gale and Greenhaven Press are registered trademarks used herein under license.

For more information, contact
Greenhaven Press
27500 Drake Rd.
Farmington Hills, MI 48331-3535
Or you can visit our Internet site at http://www.gale.com

Cover credit: © Hulton/Archive by Getty Images
Library of Congress, 25

LIBRARY OF CONGRESS CATALOGING-IN-PUBLICATION DATA

Evolution / Don Nardo, book editor.
   p. cm. — (The history of issues)
   Includes bibliographical references and index.
   ISBN 0-7377-2098-0 (lib. : alk. paper) — ISBN 0-7377-2099-9 (pbk. : alk. paper)
   1. Evolution (Biology)—History. I. Nardo, Don, 1947– . II. Series.
   QH366.2.E8459 2005
   576.8—dc22   2004047481

Printed in the United States of America

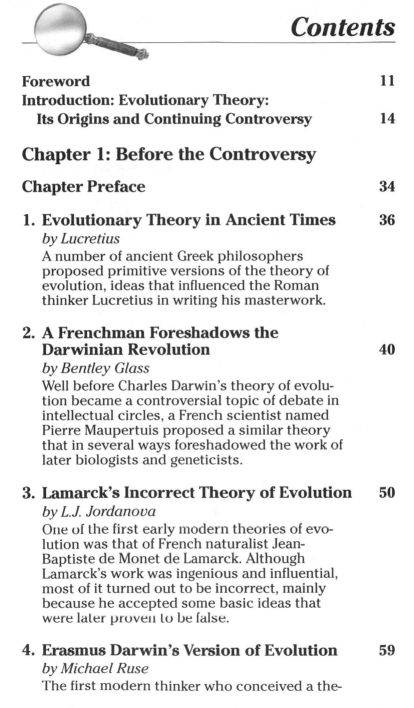

Contents

ory of evolution and related it to God and religion was Erasmus Darwin, whose grandson Charles would later rock both scientific and religious circles with his own version of evolution.

Chapter 2: Evolution and Natural Selection Create a Controversy

and therefore not nearly old enough to have supported evolutionary processes taking millions of years.

Chapter 4: Modern Developments in and Challenges to Evolution

gest that, as humans evolved, the DNA in their brains became more diverse and complex. An examination of DNA in human remains is helping to reconstruct the development of the human brain over time.

Foreword

In the 1940s, at the height of the Holocaust, Jews struggled to create a nation of their own in Palestine, a region of the Middle East that at the time was controlled by Britain. The British had placed limits on Jewish immigration to Palestine, hampering efforts to provide refuge to Jews fleeing the Holocaust. In response to this and other British policies, an underground Jewish resistance group called Irgun began carrying out terrorist attacks against British targets in Palestine, including immigration, intelligence, and police offices. Most famously, the group bombed the King David Hotel in Jerusalem, the site of a British military headquarters. Although the British were warned well in advance of the attack, they failed to evacuate the building. As a result, ninety-one people were killed (including fifteen Jews) and forty-five were injured.

Early in the twentieth century, Ireland, which had long been under British rule, was split into two countries. The south, populated mostly by Catholics, eventually achieved independence and became the Republic of Ireland. Northern Ireland, mostly Protestant, remained under British control. Catholics in both the north and south opposed British control of the north, and the Irish Republican Army (IRA) sought unification of Ireland as an independent nation. In 1969, the IRA split into two factions. A new radical wing, the Provisional IRA, was created and soon undertook numerous terrorist bombings and killings throughout Northern Ireland, the Republic of Ireland, and even in England. One of its most notorious attacks was the 1974 bombing of a Birmingham, England, bar that killed nineteen people.

In the mid-1990s, an Islamic terrorist group called al Qaeda began carrying out terrorist attacks against American targets overseas. In communications to the media, the organization listed several complaints against the United States. It generally opposed all U.S. involvement and presence in the Middle East. It particularly objected to the presence of U.S. troops in Saudi Arabia, which is the home of several Islamic holy sites. And it strongly condemned the United States for supporting the nation of Israel, which it claimed was an oppressor of Muslims. In 1998 al Qaeda's leaders issued a fatwa (a religious legal statement) calling for Muslims to kill Americans. Al Qaeda acted on this order many times—most memorably on September 11, 2001, when it attacked the World Trade Center and the Pentagon, killing nearly three thousand people.

These three groups—Irgun, the Provisional IRA, and al Qaeda—have achieved varied results. Irgun's terror campaign contributed to Britain's decision to pull out of Palestine and to support the creation of Israel in 1948. The Provisional IRA's tactics kept pressure on the British, but they also alienated many would-be supporters of independence for Northern Ireland. Al Qaeda's attacks provoked a strong U.S. military response but did not lessen America's involvement in the Middle East nor weaken its support of Israel. Despite these different results, the means and goals of these groups were similar. Although they emerged in different parts of the world during different eras and in support of different causes, all three had one thing in common: They all used clandestine violence to undermine a government they deemed oppressive or illegitimate.

The destruction of oppressive governments is not the only goal of terrorism. For example, terror is also used to minimize dissent in totalitarian regimes and to promote extreme ideologies. However, throughout history the motivations of terrorists have been remarkably similar, proving the old adage that "the more things change, the more they remain the same." Arguments for and against terrorism thus boil down to the same set of universal arguments regardless of the age: Some argue that terrorism is justified to change (or, in the case of state terror, to maintain) the prevailing political order; others respond that terrorism is inhumane and unacceptable under any circumstances. These basic views transcend time and place.

Similar fundamental arguments apply to other controversial social issues. For instance, arguments over the death penalty have always featured competing views of justice. Scholars cite biblical texts to claim that a person who takes a life must forfeit his or her life, while others cite religious doctrine to support their view that only God can take a human life. These arguments have remained essentially the same throughout the centuries. Likewise, the debate over euthanasia has persisted throughout the history of Western civilization. Supporters argue that it is compassionate to end the suffering of the dying by hastening their impending death; opponents insist that it is society's duty to make the dying as comfortable as possible as death takes its natural course.

Greenhaven Press's The History of Issues series illustrates this constancy of arguments surrounding major social issues. Each volume in the series focuses on one issue—including terrorism, the death penalty, and euthanasia—and examines how the debates have both evolved and remained essentially the same over the years. Primary documents such as newspaper articles, speeches, and government reports illuminate historical developments and offer perspectives from throughout history. Secondary sources provide overviews and commentaries from a more contemporary perspective. An introduction begins each anthology and supplies essential context and background. An annotated table of contents, chronology, and index allow for easy reference, and a bibliography and list of organizations to contact point to additional sources of information on the book's topic. With these features, The History of Issues series permits readers to glimpse both the historical and contemporary dimensions of humanity's most pressing and controversial social issues.

Introduction

Evolutionary Theory: Its Origins and Continuing Controversy

The modern debate between those who accept the theory of evolution and those who reject it in favor of the doctrine of divine creation began in earnest when Charles Darwin introduced his controversial theory of evolution in his masterwork, *The Origin of Species*, in 1859. Up until that time, the vast majority of Westerners, including almost all scientists, accepted the biblical version of creation. And even after most scientists came to accept Darwin's thesis, large numbers of people remained opposed to it. Some of these opponents eventually established the creationist movement that continues to confront, combat, and contradict evolutionary ideas. For creationists, Darwin and his theory remain enemies—misguided, sinister forces that seemingly appeared out of nowhere to challenge cherished traditional views of humanity and its place in nature.

The truth, however, is that if Darwin had not ignited the battle between evolution and creationism, someone else would have. Contrary to popular belief, he was not the first thinker or researcher to advance a theory of evolution. Indeed, his vision of plant and animal species changing slowly over time did not develop in a vacuum. Instead, his work was in some ways a continuation of a scientific dialogue that had begun many years before. During the two generations preceding the writing of Darwin's book, scientists and philosophers had been avidly discussing and debating the age of Earth, the concept of extinction, the meaning of fossils, and the doctrine of evolution itself.

In fact, several theories of evolution were advanced during this period, including two by noted French scientists and one by Darwin's own grandfather, Erasmus Darwin. What is more, even these theories were not the first of their kind. Thousands of years earlier, a number of Greek and Roman thinkers had advanced their own hypotheses about the natural progression of living things. To fully understand and appreciate how the modern controversy over evolution came about, therefore, one must examine the development of evolutionary doctrine before, as well as after, the work of Charles Darwin.

Ancient Greek and Roman Evolutionists

The first pre-Darwinian attempts to describe the evolution of living things occurred in ancient Greece. The sixth-century B.C. philosopher-scientist Anaximander was perhaps the first evolutionist (although he may well have gotten his ideas from a predecessor). He suggested that the first living creatures arose in water and that in time these creatures crawled onto the dry land and adapted themselves to their new surroundings. Moreover, he said, people developed the same way. According to Anaximander, "Man came into being from an animal other than himself, namely the fish, which in early times he resembled."[1]

In the century that followed, a fellow Greek, Empedocles, took Anaximander's theory a step further. In Empedocles' view, in the dim past, long before humans existed, numerous and diverse species of animals had existed. Some of these, he said, were not very well adapted for survival in the harsh conditions of their surroundings, so they died out. Meanwhile, stronger, more adaptable species took their place. "Monstrous and misshapen births were created," Empedocles wrote. But their existence was

> all in vain. Nature debarred them from increase [reproduction and survival]. . . . Many species must have died out altogether and failed to reproduce their kind. Every species that you now see drawing the breath of life has been protected and preserved from the beginning of the world either by cunning or by prowess or by speed.[2]

The first astonishing aspect of Empedocles' theory is that it addressed the concept of extinction, which early modern science did not begin to recognize and address until the late eighteenth and early nineteenth centuries. (It was generally assumed that all existing species had been created all at once by God and that the only species that had died out were a few that had drowned in Noah's flood.) The second remarkable aspect of Empedocles' theory was that it contained the basic concept of natural selection, or "survival of the fittest," which would later become the centerpiece of Charles Darwin's conception of evolution. What Empedocles lacked was the tremendous mass of supporting observations and other data from around the globe that Darwin was able to collect.

A few centuries after Empedocles' time, the noted Roman philosopher and poet Lucretius strongly advocated the notion of evolution. He collated the ideas of his Greek predecessors and restated them quite elegantly, including the then still little understood concept of natural selection. His panoramic vision of the upward progression of life included this insightful description of primitive human ancestors:

> Through many decades of the sun's cyclic course, they lived out their lives in the fashion of wild beasts roaming at large. No one spent his strength in guiding the curved plow. No one knew how to cleave the earth with iron, or to plant young saplings in the soil. . . . Often they stayed their hunger among the acorn-laden oaks. . . .

They did not know as yet how to enlist the aid of fire, or to make use of skins, or to clothe their bodies. . . . They lived in thickets and hillside caves and forests. . . . When night overtook them, they flung their jungle-bred limbs naked on the earth like bristly boars, and wrapped themselves round with a coverlet of leaves and branches.[3]

Trying to Discern the Age of Fossils

In the wake of Rome's fall in the fifth and sixth centuries, Europe settled into the roughly thousand-year-long period now referred to as the Middle Ages, or medieval times. In these long years, the origin of life was no longer a topic of discussion and debate, mainly because the Christian church largely shaped the thinking of the age. Church leaders strongly advocated the biblical view—namely that God had created the universe, Earth, and all the animals and plants in miraculous fashion in just a few days.

Also, the church taught, Earth was very young. Based on the evidence presented in the Bible, in fact, it could not be more than a few thousand years old. In 1650 a Protestant theologian named James Ussher carefully examined the generations of the patriarchs, priests, judges, and kings listed in the Old Testament. And he concluded that the creation had taken place in 4004 B.C. Furthermore, said Ussher, Noah's flood had occurred in the year 2349 B.C. Not long after the bishop made this pronouncement, an English scholar named John Lightfoot corroborated his calculations and published a slightly updated version. With great confidence, Lightfoot asserted that the creation commenced at exactly 9:00 A.M. on Sunday, October 23, 4004 B.C.!

In the world history described by Ussher and Lightfoot, there was not enough time for plant and animal species to change little by little and/or become extinct. The result was that few people even considered the notions of extinction and species undergoing change. Moreover, those few who did think about such things tried their best to reconcile the geologic and other physical evidence with biblical statements and calculations. For example, fossil bones that did not appear to belong to any living animal species were explained away as the remains of the unfortunate ancient beasts that God had chosen to destroy in the great flood.

As time went on, however, this explanation for strange fossils sounded less and less convincing to some scientists. They began to suspect that extinction is a natural phenomenon and that many kinds of living things had once existed and then died out. The first formal proposal of this view came in 1796. French anatomist George Cuvier presented convincing geological evidence that large-scale extinctions of species had taken place frequently in the past, and numerous other scientists agreed that this evidence was too strong to ignore.

Early Modern Evolutionary Theory

Extinction was not the only large-scale natural force that scientists began to consider during the eighteenth century. The mounting evidence for a very old Earth revived the idea of evolution, which had languished in relative obscurity since the passing of Greco-Roman civilization. The first full-blown modern theory of evolution was that of another French scientist, Pierre Maupertuis (born 1698). Maupertuis noticed that domestic plant and animal breeders routinely created new kinds of plants and animals; in his view, this meant that the existing forms of species were not completely unchangeable. Trying to explain how such changes occur, he proposed a theory of heredity that brilliantly anticipated the later discoveries of genes (the tiny particles that determine the physical characteristics of living things) and mutations (random genetic alterations that cause physical changes). "Could one not explain by that means [mutations] how from two individuals alone the multiplication of the most dissimilar species could have followed?" Maupertuis asked in 1751. "Repeated deviations would have arrived at the infinite diversity of animals that we see today . . . but to which perhaps the passage of centuries will bring only imperceptible increases."[4]

Although Maupertuis was definitely on the right track, his theory of evolution was rejected and largely forgotten by the scientific community. In part, this was because he did not have substantial observational data to back up the theory. Another, younger researcher, Erasmus Darwin, offered at least some data to support his own theory of evolution, which he expressed in piecemeal form in his *Zoonomia*, published between 1794 and 1796. Darwin cited several examples of animal species undergoing physical change, including caterpillars turning into butterflies and the creation of new breeds of dogs and horses by breeders. He also noted, as Maupertuis had, the factor of sudden mutations causing unexpected physical changes in species. The existence of these and other short-term anatomical changes suggested to Erasmus Darwin that nature might produce similar changes in the long term, that is, over thousands and millions of years. In *Zoonomia*, he asked:

> Would it not be too bold to imagine that in the great length of time since the earth began to exist . . . that all warm-blooded animals have arisen from one living filament, which The Great First Cause [God's Creation] endowed with . . . the power of acquiring new parts . . . and thus possessing the faculty of continuing to improve by its own inherent activity, and of delivering down those improvements by generation to its posterity [descendants], world without end?![5]

Erasmus Darwin's reference to God in this passage is crucial. Though he believed that evolution was a fact, it did not diminish his belief in a divine creator. In his view, God's singular and magnificent achievement had been to set the process of evolution in motion. Thus, Darwin embodied a unique coalescence of evolutionary science and Christian creationism, a worldview in which each doctrine easily reconciled with the other. Had all, or at least most, scientists, theologians, and ordinary people of that era reached this same personal reconciliation, the great controversy and still ongoing battle between evolutionists and creationists would not have materialized.

Darwin and Natural Selection

As it was, however, that controversy did come to pass. It was not Erasmus Darwin's book and ideas that sparked it, however. Nor was it the work of another important early evolutionist, French naturalist Jean-Baptiste de Monet de Lamarck (born 1744). Lamarck's theory, usually referred to as transformism, was roughly similar to Darwin's except that Lamarck proposed a different motive cause for changes in species. Darwin pointed to a fierce competition for food and other resources; in this competition, he said, the strongest and most adaptable individuals and varieties survived while the others died out. Lamarck took a different approach. He held that each new generation of a species inherited "acquired characteristics." Changes in these characteristics could accumulate over time and thereby transform the species. A well-known example he used was successive generations of giraffes, each of which attempted to reach higher tree leaves; over time, their necks got longer and longer until the breed had undergone enough physical change to warrant the status of a new species.

Modern scientists eventually showed that Lamarck was wrong. The rise of the science of genetics demonstrated that the genes, which carry the genetic code, do not change as a result of what an animal does or experiences in its life. Lamarck's work was important, however, because it inspired a number of other European scientists to examine the notion of evolution.

The most famous and influential of these men, of course, was Erasmus Darwin's grandson, Charles, who was born in a small town in western England in 1809. It was not until the 1830s that the younger Darwin began to accumulate the experiences, discoveries, observations, and clues that would eventually lead him to his masterful vision of the evolution of life on Earth. In 1831 he took a job as the official naturalist aboard the HMS *Beagle*, which set out on a five-year voyage of scientific exploration. The expedition visited distant, little-known continents and islands, including the remote and primitive Galápagos Islands (about five hundred miles off the western coast of South America). There, Darwin became fascinated by numerous strange, unexplainable variations in some of the animal species he observed. The desire to understand why these variations had come to exist piqued his interest in the idea of evolution and inspired him to begin accumulating the enormous quantities of evidence he would eventually cite in his masterwork, *The Origin of Species*.

The main thrust of the book, which was published in November 1859, was a long, complex presentation of Darwin's mechanism for evolutionary change—natural selection, which his grandfather had touched on in a more cursory way. In Charles Darwin's more developed version, physical characteristics such as size, strength, shape of body parts, and quality of vision and hearing, regularly pass from parents to offspring. This process is random, however, and always results in tiny variations from one generation to another. Also, life consists of a fierce struggle for existence in which all species compete for the same limited supplies of food, water, and territory.

Working through, or taking advantage of, these generational variations and the struggle for existence, Darwin wrote, nature "selects," or allows the survival of, those individuals whose variations are favored over those of others. In other words, plants and animals that manage to adapt to changing environmental conditions will survive and pass on their favorable characteristics to their offspring. These new, favored kinds of living things will, over time, become increasingly different from their parents and, after the passage of thousands of generations, they will have become so different that they can be categorized as a new species. At the same time, species with favored characteristics tend to crowd out those that cannot adapt as quickly or as well. These less successful living things inevitably become extinct. Although the process of evolution proceeds at much too slow a pace to be directly seen, Darwin, explained, its workings can be seen in the fossil record, which reveals the rise and fall of countless species over the eons. Darwin's thesis neatly explained why the most ancient fossils least resemble modern ones. Succeeding generations of offspring become increasingly less like an original set of parents and the longer evolution proceeds, the less the older forms resemble the new ones.

Darwin Gains Both Supporters and Enemies

The concept that all living things have been and remain locked in an eternal, violent struggle in which the strong survive and the weak die out disturbed many of the people who read or heard about Darwin's book. To them, the theory seemed to go against the traditional, widely accepted view that God had miraculously created all the plant and animal species, whose forms were unchanging. It is not surprising, therefore, that the book stirred up controversy and provoked several prominent scholars and clergymen to denounce it and its author. One noted churchman called Darwin the most dangerous man in England. And the respected geologist Adam Sedgwick, who had once been Darwin's teacher and friend, described whole sections of the book as false and troubling. Darwin's theory, said Sedgwick, thrust the human race into a degrading position no better than that of wild beasts.

It was at this point—the publication of Darwin's *Origin*—therefore, that the great rift between modern evolutionists and creationists began to form. It is interesting to note in retrospect that Darwin's supposed degradation of the human race appears nowhere in his book. In fact, *Origin* does not deal directly with human evolution from earlier primates; Darwin addressed that topic in his 1871 book, *The Descent of Man*. And in neither book did he suggest that his theory had anything to do with God or the biblical view of creation. "I see no good reason why the views given in this volume should shock the religious feelings of anyone," he wrote in the conclusion of *Origin.*

> A celebrated author . . . has written to me that "he has gradually learned to see that it is just as noble a conception of the Deity [i.e., God] to believe that He created a few original forms capable of self-development into other and needful forms, as to believe that He required a fresh act of creation to supply the voids caused by His laws."[6]

This was the same theme of reconciling science and religion that Erasmus Darwin had championed two generations earlier. And in the wake of the sensation over *Origin*, many scientists and religious leaders alike began to accept this moderate approach. In less than a decade, nearly all reputable scientists had come to accept Darwin's version of evolution either fully or in part. One result of this trend was that standard biology textbooks began to include sections on evolution. And those that did not, rapidly became outdated. By 1920, some sixty years after Darwin's book had appeared, nearly every college and high school in both Europe and the United States routinely taught evolution in biology classes.

In the latter decades of the nineteenth century several respected biblical scholars were in the midst of new, intense studies of early versions of the Bible as well as the cultures of the peoples who lived in the ancient lands described in its pages. These studies showed that the Old Testament had been produced over a span of several hundred years by a variety of authors. In addition, a number of biblical stories had been based on or influenced by the legends of Near Eastern peoples such as the Babylonians. This implied that at least some sections of the Bible must consist of fable or allegory. And this led many religious leaders to attempt to reconcile the Bible with new and widely accepted scientific discoveries, including Darwin's vision of evolution.

Not all of the faithful were prepared for such a reconciliation, however. During the late nineteenth and early twentieth centuries, most fundamentalists, primarily Americans who lived in the rural south (the so-called Bible Belt), refused to accept any theory of evolution, whether Darwin's or someone else's. The name of their group derived from a series of ten pamphlets, collectively called *The Fundamentals*, that were published in 1910. These writings, which circulated throughout the United States in the following years, were a reaction by very conservative biblical literalists to what they saw as a steady erosion of traditional faith in the Bible in favor of "godless" science. *The Fundamentals* tried to redefine what it meant to be a Christian, stressing the "Five Points" of true belief. These were the complete infallibility of the Bible; the virgin birth of Jesus Christ; Christ's voluntary death to atone for humanity's sins; Christ's resurrection into heaven; and the authenticity of all miracles described in the Bible.

Creationism vs. Evolution

Although these very conservative concepts remained outside of the intellectual mainstream of the major organized religions, they strongly appealed to a growing number of people. Consequently, the fundamentalist movement rapidly gained strength in the United States and Canada. And by the early 1920s it had achieved a prominent position in American religion and society.

The fundamentalists did not merely oppose the concept of evolution. They were (and still are) particularly disturbed by the systematic teaching of that doctrine in schools since the biblical version of the origins of life was usually excluded from classroom biology lessons. It therefore became a major mission of the fundamentalist movement to rectify this situation. The group's early efforts in this regard were unsuccessful. It was not long, however, before it scored some modest victories. In 1923 Oklahoma passed a bill that banned textbooks containing descriptions of evolution (though the actual teaching of evolution remained legal).

Clarence Darrow (leaning on table) defends John Scopes during the Scopes "Monkey Trial" of 1925. Scopes was prosecuted for teaching evolution.

The fundamentalists' most widely publicized effort against the teaching of evolution occurred in the infamous Scopes "Monkey Trial" in 1925. On May 7 John Scopes, a high school science teacher in Dayton, Tennessee, was arrested for breaking a newly passed state law that banned the teaching of evolution. The trial, which was held in Dayton from July 10 to July 21, 1925, attracted large crowds and global press coverage. Although the jury found Scopes guilty and ordered him to pay a fine, national public opinion viewed the trial as an intellectual and moral victory for the defense.

The fundamentalists, many of whom now called themselves creationists, did not see themselves as defeated by the Scopes incident, however. In the decades following the trial, they kept up their crusade to remove or at least water down the teaching of evolution in schools. A law similar to the one in Tennessee passed in Mississippi in 1926, for instance. In time, however, this approach to achieving creationist goals proved fruitless. In 1968 the U.S. Supreme Court ruled (in the *Epperson vs. Arkansas* case) that such laws were unconstitutional because they violate the First Amendment's clause stipulating that church and state be separate.

Still undaunted, the creationists changed their overall approach and began arguing that their vision of the creation constitutes a viable alternative scientific theory. This theory, they maintained, is as worthy as evolution and should be taught right alongside it in biology classes. This new approach brought some moderate successes. In 1981, for example, the Arkansas legislature passed Act 590, which stipulated that biology teachers had to cover "creation science" along with "evolution science" in their courses. The law never really went into effect, though, because numerous groups and individuals opposed it and filed suit to nullify it.

Despite this and a few other similar setbacks, some creationists kept up the fight against evolution. In the 1990s they began to stress that evolution is not only wrong, but it is also not scientific. In this view, evolution is just another belief system like creationism (since no one was around to witness the creation and verify how it was accomplished). Therefore, both belief systems should receive an equal hearing in classrooms. At least, they said, if evolution must remain in schools, it should not be presented as an accepted scientific fact. Following this reasoning, in 1995 the Alabama state board of education passed a proposal calling for biology textbooks to include a disclaimer saying that evolution is a controversial theory that only some scientists accept. And in 1998 Washington State's legislature considered a bill along these same lines.

Distortions of Darwinian Evolution

These and other clashes between modern scientists and educators and the creationist movement demonstrate that the controversy over evolution was far from dead during the twentieth century. Another dimension of that controversy revolved around the attempted application of Darwin's ideas to nonscientific areas, which almost always constituted gross distortions of his views. The most familiar examples are often lumped together under the generalized term *social Darwinism*. Essentially, social Darwinists were (and remain) people who tried to interpret and apply Darwin's concept of the natural selection of species to human moral, social, and economic contexts. Darwin himself witnessed the beginnings of this reinterpretation of his ideas. He tried to make light of it in an 1860 letter to a colleague, joking, "I have received in a Manchester newspaper rather a good squib [review], showing that I have proved that might is right and therefore that Napoleon is right, and that every cheating tradesman is also right."[7]

A majority of scholars believe that the father of social Darwinism was a contemporary of Darwin's—the British sociologist and philosopher Herbert Spencer. In fact, it was Spencer, not Darwin, who coined the term *survival of the fittest*, which is often used synonymously with natural selection. In Spencer's view, all evil and suffering in human society is the result of the failure of people to adapt to their changing environment, especially changes within their societies. Human society was once primarily animalistic and brutal, he contended. Improvements have been made over time; however, humanity's achievement of a peaceful social existence is still incomplete. Aggression and violence, Spencer said, have yet to be totally bred out of the human species. Therefore, the best course for society is to punish dishonesty and laziness and to reward honesty, diligence, and cooperation. That way, human society will continue to better itself and progress as time goes on.

Eventually, scientists showed that Spencer's social extension of Darwin's views amounted to an artificial construct based more on wishful thinking than facts. But at least Spencer's misapplication of science can be forgiven for its largely benign approach to improving society. The same cannot be said for two other twentieth-century distortions of Darwin's ideas—the racial application of social Darwinism and the eugenics movement. The racial aspect of social Darwinism was spawned mainly by the rise of the science of genetics in the early 1900s. A number of people—both scientists and nonscientists—were impressed by the notion that human physical and mental attributes were determined and controlled to some unknown degree by genes within the body's cells. This led them to use genetic theory to prove or justify their personal likes and dislikes for peoples and groups who were different. In this view, some races and groups were genetically inferior to others because their genes were inferior or defective.

Unfortunately, this twisted interpretation of science led a number of governments to apply the doctrine of survival of the fittest in an attempt to "racially purify" their societies. The most infamous example of this eugenics movement was, of course, Nazi dictator Adolf Hitler's extermination of millions of Jews, Slavs, Gypsies, and others whom he saw as inferior and undesirable. In Hitler's view, only the fittest were worthy of surviving. And the fittest were the German "Aryans," the fair-skinned race that had been destined by history to rule the world. Scientists have shown that no such pure Aryan race exists or has ever existed. But Hitler was convinced of its reality and warned that Aryans must not make the mistake of diluting their blood through intermarriage with inferior peoples. "Blood mixture and the resultant drop in the racial level," he wrote,

> is the sole cause of the dying out of old cultures; for men do not perish as a result of lost wars, but by the loss of that force of resistance which is contained only in pure blood. All who are not of good race in this world are chaff [refuse or waste material].[8]

Many people today are unaware that Hitler's crimes against humanity were only one part of a much larger eugenics movement that swept western Europe, Canada, and the United States in the early decades of the twentieth century. Laws were passed in a number of U.S. states, for example, allowing criminals, retarded people, and the mentally ill to be sterilized. This, it was thought, would improve society's overall racial stock. In a similar vein, other laws prohibited white people from marrying black people, who were seen by the white authorities as inferior.

Further, in 1924, Congress passed the Immigration Restriction Act, designed to reduce the number of immigrants entering the United States from eastern and southern Europe. Behind this legislation was the belief that Italians, Greeks, Poles, Jews, and other groups were racially inferior to Americans of British and German descent. Incredibly, despite the fact that scientists completely discredited these ideas in the 1930s and 1940s, the Immigration Restriction Act was not repealed until the 1960s.

Twentieth-Century Evolutionary Theory

Fortunately, the continuing march of science did manage to stifle the eugenics movement and other such distortions of legitimate evolutionary concepts. At the same time, twentieth-century science made a number of important advances that carried evolutionary theory in new and sometimes unexpected directions. Darwin's *Origin of Species* had been a milestone of science that had convinced the vast majority of scientists that evolution was a real process. However, not all scientists were ready to accept that the principle of natural selection was the main driving force behind evolution, as Darwin had contended. In the late nineteenth and early twentieth centuries, therefore, some researchers explored a number of alternative hypotheses for the mechanism directing evolution.

Seemingly very promising in this regard was the emergence of the science of genetics in the early 1900s. Researchers began to suspect that not only are physical characteristics determined by genes but also that mutations, or changes in the genes, might, over time, alter an entire species. As time went on, scientists learned that large-scale mutations are harmful and that smaller ones are responsible for a certain portion of positive evolutionary change. Moreover, mutation itself is not the chief source of variations in the genes. Instead, a complex recombination of the genetic material produces the majority of the change in both individuals and species.

Of course, there still had to be some way for nature to choose which genetic combinations would be successful and which would not. This is where Darwin's theory of natural selection came to the fore. Scientists showed that those individuals in a given population of animals whose gene combinations allow them to adapt well to their environment are the ones who are most likely to make the species survive. Thus, natural selection controls the flow of genes through a population of animals or plants.

In this and similar ways, Darwin's theory of natural selection and the new science of genetics achieved a sort of synthesis, or melding together, in the 1930s and 1940s. Accordingly, scientists came to call it the evolutionary synthesis (or the modern evolutionary synthesis, or simply the modern synthesis). Important figures in developing this new way of looking at evolution were Thomas Hunt Morgan, who made key strides in understanding mutation; Ernst Mayr, who organized ideas from many disparate scientific disciplines into a single system of thought; and Theodosius Dobzhansky, Julian Huxley, August Weismann, George Gaylord Simpson, and G.L. Stebbins, all of whom made important individual contributions.

These pioneers did not have the last say in the modern understanding of evolution, however. In the years that followed, other scientists proposed new evolutionary ideas that either further refined or to some degree challenged the general notions of the evolutionary synthesis. A crucial factor that made these newer theories possible was the emergence of an understanding of DNA, the chemical that directs heredity in the genes. Scientists had suspected the existence of such a key chemical for a long time. As early as 1944 physicist Erwin Schrödinger hypothesized that a "hereditary code-script" was involved in reproduction and evolution. But he and others were unable to isolate the elusive substance or figure out exactly how it worked.

Then, in 1953, James Watson and Francis Crick stunned the scientific world with their discovery of the structure and workings of the DNA molecule. Understanding of DNA has given scientists an increased understanding of the workings of such evolutionary processes as genetic drift (slight variations of genetic traits within a given population of animals) and gene flow (or gene migration, the flow of genes from one population to another via interbreeding).

Darwin did not foresee DNA. Nor did he imagine the rash of new theories that came in the wake of the evolutionary synthesis and discovery of DNA. In the 1970s scientists Stephen Jay Gould and Niles Eldredge proposed that evolution sometimes acts in quick spurts rather than gradually; this idea became known as punctuated equilibrium. Soon afterward, researcher Lynn Margulis and others developed the theory of symbiogenesis. It suggests that evolution occurs on the molecular level when genes from different organisms (deriving originally from the genetic material of bacteria) merge. Others, including Michael J. Behe, have recently argued the opposite—that little or no evolution is occurring on the molecular level.

These and other evolutionary theories and ideas, as well as ongoing battles between scientists and creationists, make one thing clear to all those who either accept or reject the notion of evolution. Namely, the controversy stirred up by Darwin and his supporters in the nineteenth century has not abated and is likely to continue well into the future. It is probable that Darwin himself would find most of these arguments and at times legal disputes disquieting and distasteful since he consistently desired to avoid controversy. (Ironically, of course, his masterwork stirred up enormous controversy, much to his displeasure.) In his autobiography, compiled mostly in 1876, he wrote:

> My views have often been grossly misrepresented, bitterly opposed and ridiculed. . . . [And] on the whole I do not doubt that my works have been over and over again greatly over-praised. . . . [A close colleague] strongly advised me never to get entangled in controversy, as it rarely did any good and caused a miserable loss of time and temper.[9]

Notes

1. Quoted in Philip Wheelwright, ed., *The Presocratics.* New York: Macmillan, 1966, p. 58.

2. Quoted in Lucretius, *On Nature*, published as *Lucretius on the Nature of the Universe*, trans. Ronald Latham. Baltimore: Penguin, 1951, p. 197.

3. Lucretius, *On Nature*, pp. 199–200.

4. Pierre Maupertuis, *Nature's System*, in *Forerunners of Darwin: 1745–1859*, ed. Bentley Glass et al. Baltimore: Johns Hopkins University Press, 1968, p. 77.

5. Quoted in Desmond King-Hele, *Erasmus Darwin.* New York: Scribner's, 1963, p. 69.

6. Charles Darwin, *The Origin of Species.* New York: New American Library, 1958, p. 443.

7. Quoted in Peter Singer, *A Darwinian Left: Politics, Evolution, and Cooperation.* New Haven, CT: Yale University Press, 1999, p. 10.

8. Adolf Hitler, *Mein Kampf*, trans. Ralph Manheim. Boston: Houghton Mifflin, 1971, p. 296.

9. Charles Darwin, *Autobiography and Selected Letters*, ed. Francis Darwin. New York: Dover, 1958, p. 46.

Before the Controversy

Chapter Preface

I t is reasonable to ask why the theory of evolution did not
create any serious controversy before Charles Darwin
published his highly influential *Origin of Species* in 1859. After all, a number of scientists and philosophers had earlier
proposed their own evolutionary ideas. These included Darwin's own grandfather, Erasmus Darwin, some noted
eighteenth-century French researchers, and several ancient
Greek and Roman thinkers. Some of these men even discussed the concept of "survival of the fittest" (in which the
strongest and most adaptable species compete with and
crowd out many weaker, less adaptable ones), although
these thinkers did not offer the substantial amounts of observational evidence that Charles Darwin did.

One reason that these earlier evolutionary theories did
not stir up any significant controversy was the low degree
of literacy and/or interest in scientific matters among average people. All through ancient and medieval times and
well into early modern times, the majority of people could
not read (or did not read very well). Also, few had access
to books. And very few took part in discussions about scientific ideas or were even exposed to such concepts. In ancient times, for instance, there was no mass media, no pub-
lishing industry, no bookstores, and no public schools or
universal education. Only rarely did new scientific ideas fil-
ter down to the lower classes, whose members made up
the bulk of the population. So average persons were more
likely to accept with little question their society's long-held
traditional explanations for the creation, the generation of
living things, and other aspects of the natural world.

Religious tolerance was another important reason that
evolution did not cause any appreciable controversy in an-

cient times. One of the chief reasons that Charles Darwin's theory upset so many people when it appeared (and still bothers some people) is that they felt such ideas threatened their cherished religious beliefs. In their view, to accept Darwin's ideas was to deny the traditional creation story of Christianity as set down in the Bible. In the ancient Greco-Roman world, in contrast, few people felt threatened by others' religious or intellectual beliefs. A wide range of faiths and gods proliferated across the Roman Empire; and all were accepted as viable alternative paths to the same eternal truths. (The ancient Jews and Christians were exceptions. They were often persecuted, but not for their religious beliefs. Rather, their stubborn exclusivity, insistence that only their own god existed, and refusal to accept the viability of other people's gods and beliefs got them into trouble.)

By Charles Darwin's time, the European world had changed a great deal. Large numbers of people could read. And though most did not read Darwin's book or other scientific works, they read about it in newspapers or heard about it in churches, the workplace, or elsewhere. Also, there was very little religious diversity in Darwin's society. Most people were (and remain today) devout Christians; their feelings of religious exclusivity remained strong, and many were suspicious of new ideas that differed from those they had learned as children in Sunday school. In such an atmosphere, a quiet discussion among a few scientists in a private setting was acceptable; but a public dialogue that caused a sudden change in the way people viewed their origins was disturbing and much less acceptable. For these and other reasons, Darwin's book made evolution a controversial topic for the first time in history.

Evolutionary Theory in Ancient Times

LUCRETIUS

The concept of species evolving into new and different phys-
ical forms over time was not new to modern science. Several
ancient Greek philosopher-scientists, including Anaximander
and Empedocles, considered the idea that humans and other
complex animals sprang from lower animal forms such as
fish. The first-century B.C. Roman philosopher and poet Lu-
cretius echoed these ideas in his lengthy treatise On Nature,
in which he advocated not only that matter is made up of
atoms (an idea he got from Democritus, Epicurus, and other
Greek thinkers), but also that some sort of natural progres-
sion had produced the diversity of life-forms observed in his
day. (It is unclear how much of Lucretius's text was based on
now lost copies of the works of Empedocles and other Greeks
and how much was original to Lucretius himself; most histo-
rians think he was not particularly original.) In the following
excerpt from On Nature, *Lucretius emphasizes that "trans-*
formations" of various animals are natural and have hap-
pened over long periods of time; he also addresses the key
concept of extinction. These are basic biological tenets ac-
cepted by modern scientists.

The name of mother has rightly been bestowed on the
earth, since it brought forth the human race and gave
birth at the appointed season to every beast that runs wild

Lucretius, *Lucretius: On the Nature of the Universe*, translated by Ronald
Latham. Baltimore, MD: Penguin, 1962.

among the high hills and at the same time to the birds of the air in all their rich variety.

Then, because there must be an end to such parturition, the earth ceased to bear, like a woman worn out with age. For the nature of the world as a whole is altered by age. Everything must pass through successive phases. Nothing remains for ever what it was. Everything is on the move. Everything is transformed by nature and forced into new paths. One thing, withered by time, decays and dwindles. Another emerges from ignominy, and waxes strong. So the nature of the world as a whole is altered by age. The earth passes through successive phases, so that it can no longer bear what it could, and it can now what it could not before.

Nature Weeded Out the "Monsters"

In those days the earth attempted also to produce a host of monsters, grotesque in build and aspect—hermaphrodites, halfway between the sexes yet cut off from either, creatures bereft of feet or dispossessed of hands, dumb, mouthless brutes, or eyeless and blind, or disabled by the adhesion of their limbs to the trunk, so that they could neither do anything nor go anywhere nor keep out of harm's way nor take what they needed. These and other such *monstrous and misshapen births were created. But all in vain.* Nature debarred them from increase. They could not gain the coveted flower of maturity nor procure food nor be coupled by the arts of Venus. For it is evident that many contributory factors are essential to the reproduction of a species. First, it must have a food-supply. Then it must have some channel by which the procreative seeds can travel outward through the body when the limbs are relaxed. Then, in order that male and female may couple, they must have some means of interchanging their mutual delight.

In those days, again, *many species must hare died out altogether* and failed to reproduce their kind. Every species that you now see drawing the breath of life has been protected and preserved from the beginning of the world either by cunning or by prowess or by speed. [This is an early, very rudimentary statement of the doctrine of natural selection, or the "survival of the fittest."] In addition, there are many that survive under human protection because their usefulness has commended them to our care. The surly breed of lions, for instance, in their native ferocity have been preserved by prowess, the fox by cunning and the stag by flight. The dog, whose loyal heart is alert even in sleep, all beasts of burden of whatever breed, fleecy sheep and horned cattle, over all these, my Memmius, man has established his protectorate. They have gladly escaped from predatory beasts and sought peace and the lavish meals, procured by no effort of theirs, with which we recompense their service. But those that were gifted with none of these natural assets, unable either to live on their own resources or to make any contribution to human welfare, in return for which we might let their race feed in safety under our guardianship—all these, trapped in the toils of their own destiny, were fair game and an easy prey for others, till nature brought their race to extinction.

Species Develop According to Natural Law

But *there never were*, nor ever can be, Centaurs—*creatures with a double nature*, combining organs of different origin in a single body so that there may be a balance of power between attributes drawn from two distinct sources. This can be inferred by the dullest wit from these facts. First, a horse reaches its vigorous prime in about three years, a boy far from it: for often even at that age he will fumble in sleep for his mother's suckling breasts. Then, when the horse's limbs are flagging and his mettle is fading with the onset of age and the ebbing of life, then is the very time when the boy is crowned with the flower of youth and his cheeks are clothed with a downy bloom. You need not suppose, therefore, that there can ever be a Centaur, compounded of man and draught-horse, or a Scylla, half sea-monster, with a girdle of mad dogs, or any other such monstrous hybrid between species whose bodies are obviously incompatible. They do not match in their maturing, in gaining strength or in losing it with advancing years. They respond diversely to the flame of Venus. Their habits are discordant. Their senses are not gratified by the same stimuli. You may even see bearded goats battening on hemlock, which to man is deadly poison. Since flame sears and burns the tawny frames of lions no less than any other form of flesh and blood that exists on earth, how could there be a Chimaera with three bodies rolled into one, in front a lion, at the rear a serpent, in the middle the she-goat that her name implies, belching from her jaws a dire flame born of her body? If anyone pretends that such monsters could have been begotten when earth was young and the sky new, pinning his faith merely on that empty word 'young', he is welcome to trot out a string of fairy tales of the same stamp. Let him declare that rivers of gold in those days flowed in profusion over the earth; that the trees bore gems for blossoms, or that a man was born with such a stretch of limbs that he could bestride the high seas and spin the whole firmament around him with his hands. The fact that there were abundant seeds of things in the earth at the time when it first gave birth to living creatures is no indication that beasts could have been created of intermingled shapes with limbs compounded from different species. The growths that even now spring profusely from the soil—the varieties of herbs and cereals and lusty trees—cannot be produced in this composite fashion: each species develops according to its own kind, and they all guard their specific characters in obedience to the laws of nature [a concept foreshadowing that of genes and heredity which were not demonstrated as fact until early modern times].

A Frenchman Foreshadows the Darwinian Revolution

BENTLEY GLASS

Most people today are unaware that Charles Darwin was not the first modern scientist to propose a theory of the evolution of plants and animals. That distinction belongs to the brilliant eighteenth-century French scientist-philosopher Pierre Maupertuis. For Maupertuis, an important initial clue to the existence of the evolutionary process was the manner in which domestic plant and animal breeders created new hybrid types of animals and plants. Their work suggested to him that under certain conditions the forms of species were not fixed, but could change. He also observed the phenomenon that later came to be called mutation, and correctly noted its importance in the generation of new plant and animal forms. Maupertuis and his ideas were largely forgotten probably at least in part because he lacked what Charles Darwin provided later—a mass of observational data collected over several decades to back up the revolutionary tenets of his theory. This overview of Maupertuis's ideas and work is by science historian Bentley Glass.

Bentley Glass, "Maupertuis, Pioneer of Genetics and Evolution," *Forerunners of Darwin: 1745–1859,* edited by Bentley Glass, Owsei Temkin, and William L. Straus Jr. Baltimore, MD: The Johns Hopkins University Press, 1968. Copyright © 1959 by The Johns Hopkins University Press. All rights reserved. Reproduced by permission of The Johns Hopkins University Press.

The mid–eighteenth century was a period almost unexampled in the vigor and advancement of science. Newton's physics had finished remaking the Heaven and the Earth. In biology, the new classification of plants and animals, introduced in 1735 by Linnaeus and developed into the binomial system in 1753, gave a vast stimulus to the discovery and description of new species. Louis Leclerc, Comte de Buffon, was beginning his tremendous *Natural History*, which ran to 36 volumes, yet was a best-seller to be found in the library of every European with any pretension to culture. . . .

An Early Champion of Newton

Preeminent among natural philosophers of the time was Pierre Louis Moreau de Maupertuis, President of Frederick the Great's Academy of Sciences in Berlin. As a young man, Maupertuis was the first person on the Continent to understand and appreciate Newton's laws of gravitation; and indeed, it was through Maupertuis that [the renowned French writer François-Marie] Voltaire first became convinced of their truth. When thirty years of age, in 1728, Maupertuis had visited London. It was the year of Newton's death. Here Maupertuis became a member of the Royal Society and a disciple of Newton. Upon his return to France, at a time when Newton's theory of gravitation was still violently opposed by [a majority of scientists] . . . Maupertuis became the open defender and expounder of the new scientific doctrines, just as [Thomas H.] Huxley over a century later sprang to the defense of Darwinism. . . . In that same year Voltaire, more and more interested in scientific pursuits . . . wrote to Maupertuis, asking him in flattering terms for his judgment upon Newton's theory. Voltaire's conversion followed, and later, his own work on the subject, which was by far his most serious scientific production.

Four years later, in 1736, Maupertuis headed one of two expeditions sent out to test the flattening of the earth toward the poles, by accurately measuring a degree along a meridian of longitude in two places, in the one instance at the equator, and in the other, just as far to the north as feasible. Maupertuis directed the expedition into Lapland, [French soldier and geographer Charles Marie de] La Condamine the expedition to Peru; and between them they provided the first convincing proof to the world that Newton was correct, just as the expeditions to measure the gravitational deflection of light rays around the sun during an eclipse were in a later day and age to confirm the relativity theory of [noted physicist Albert] Einstein. There was no Nobel prize in those days, or surely Maupertuis would have earned one, at the age of 38.

Upon his return from Lapland, Maupertuis was addressed by Voltaire in these flattering words (letter of 19th Jan., 1741): "M. Algarotti is count; but you—you are marquis of the arctic circle, and you have won for yourself one degree of the meridian in France and one in Lapland. Your name covers a good part of the globe. I find you really a very great seigneur. Remember me in your glory." It was, in fact, on Voltaire's recommendation that, in 1740, Maupertuis was invited by Frederick the Great to come to Berlin as head of the reorganized Academy of Sciences.

Maupertuis, however, came to consider that his work as a champion of Newton on the Continent afforded him no great personal glory, and therefore in later years he laid great weight on his discovery in 1746 of the Principle of Least Action, which is all too commonly credited to one of the three great mathematicians, Euler, Lagrange, and Hamilton, who further developed it. This Principle is indeed one of the greatest generalizations in all physical science, although not fully appreciated until the advent of quantum mechanics in the [twentieth] century. . . .

Genes and Physical Traits

Eminent as were these contributions to physical science and to philosophy, it is in his biological ideas that Maupertuis was most clearly gifted with prevision. Here he must be reckoned as fully a century or a century and a half before his time. His biological ideas may be considered under the three heads of the formation of the individual, the nature of heredity, and the evolution of species, although obviously these are so closely interrelated that the division is largely artificial.

He began with an interest in the formation of the embryo, and quickly put his finger on the soundest argument against the pre-formationists, who believed the embryo to be fully preformed before conception, and present either in the sperm or the egg. This argument led him to a study of heredity, and he may be justly claimed as the first person to record and interpret the inheritance of a human trait through several generations. He was also the first to apply the laws of probability to the study of heredity. He was led by the facts he had uncovered to develop a theory of heredity that astonishingly forecast the theory of the genes. He believed that heredity must be due to particles derived both from the mother and from the father, that similar particles have an affinity for each other that makes them pair, and that for each such pair either the particle from the mother or the one from the father may dominate over the other, so that a trait may seemingly be inherited from distant ancestors by passing through parents who are unaffected. From an accidental deficiency of certain particles there might arise embryos with certain parts missing, and from an excess of certain particles could come embryos with extra parts, like the six-fingered persons or the giant with an extra lumbar vertebra whom Maupertuis studied. There might even be complete alterations of particles— what today we would call "mutations"—and these fortuitous changes might be the beginning of new species, if acted upon by a survival of the fittest and if geographically isolated so as to prevent their intermingling with the original forms. In short, virtually every idea of the Mendelian [relating to Gregor Mendel, the Austrian botanist who founded the science of genetics] mechanism of heredity and the classical Darwinian reasoning from natural selection and geographic isolation is here combined, together with [Dutch botanist] De Vries' theory of mutations as the origin of species, in a synthesis of such genius that it is not surprising that no contemporary of its author had a true appreciation of it. . . .

A House Filled with Animals

But Maupertuis was not content to make only this analysis. He undertook actual breeding experiments with animals to test out his theories, although of the results of these he has unfortunately left us only the account of a single one. It is related that he "adored animals and lived surrounded by them." "You are more pleased with Mme. d'Aiguillon than with me," wrote Mme. du Deffand to him one day, "she sends you cats." And Frederick wrote, too: "I know that at Paris just as at Berlin you are enjoying the delights of good company. . . . I am only afraid that Mme. la duchesse d'Aiguillon is spoiling you. She loves parrots and cats, which, is a prodigious merit in your eyes. . . ." Maupertuis had established himself in the outskirts of Berlin, in a spacious house adjacent to the royal park, near the present Tiergarten; and this house he had converted into a virtual Noah's ark. Samuel Formey, permanent secretary of the Berlin Academy, has left us the following description: "The house of M. de Maupertuis was a veritable menagerie, filled with animals of every species, who failed to maintain the proprieties. In the living-rooms troops of dogs and cats, parrots and parakeets, etc. In the fore-court all sorts of strange birds. He once had sent from Hamburg a shipment of rare hens with a cock. It was sometimes dangerous to pass by the run of these animals, by whom some had been attacked. I was especially afraid of the Iceland dogs. M. de Maupertuis amused himself above all by creating new species by mating different races together; and he showed with complaisance the products of these matings, who partook of the qualities of the males and of the females who had engendered them. I loved better to see the birds, and especially the parakeets, which were charming. . . ."

It was of the Iceland dogs that Maupertuis has left us the account of his breeding experiment: "Chance led me to meet with a very singular bitch, of that breed (*espèce*) that is called in Berlin the Iceland Dogs: she had her whole body the color of slate, and her head entirely yellow; a singularity which those who observe the manner in which the colors are distributed in this sort (*genre*) of animals will find perhaps rarer than that of supernumerary digits. I wished to perpetuate it; and after three litters of dogs by different fathers which did not yield anything of the sort, at the fourth litter she gave birth to one who possessed it. The mother died; and from that dog, after several matings with different bitches, there was born another who was exactly like him. I actually have them both." His breeding of dogs led him to wonder particularly about the supernumerary fifth digit which is not uncommon on the hind foot: "There are no animals at all upon whom supernumerary digits appear more frequently than upon dogs. It is a remarkable thing that they ordinarily have one digit less on the hind feet than on those in front, where they have five. However, it is not at all rare to find dogs who have a fifth digit on the hind feet, although most often detached from the bone and without articulation. Is this fifth digit of the hind feet then a supernumerary? or is it, in the regular course, only a digit lost from breed (*race*) to breed throughout the entire species, and which tends from time to time to reappear? For mutilations can become hereditary just as much as superfluities." Were all dogs, in other words, once five-toed on both front and back feet? Have we here a remnant, a vestige of a once functional structure? These observations might well have been made by Charles Darwin.

Maupertuis's Theory of Evolution

Maupertuis's studies thus led him to evolution. Here with certainty he must be ranked above all the precursors of

Darwin. To begin with, he was faced with the problem of accounting for supernumerary digits, albinism [lack of normal skin pigmentation], and other hereditary anomalies on the basis of his theory of generation. This he solved ingeniously. "If each particle is united to those that are to be its neighbors, and only to those, the child is born perfect. If some particles are too distant, or of a form too little suitable, or too weak in affinity to unite with those with which they should be united, there is born a monster with deficiency (*monstre par défaut*). But if it happens that superfluous particles nevertheless find their place, and unite with the particles whose union was already sufficient, there is a monster with extra parts (*monstre par exces*)." Even Mendel did not foresee that deficiencies and duplications of the hereditary material might constitute a basis of abnormal development, a sort of mutation! Maupertuis comments on the remarkable fact that in monsters with extra parts, these are always to be found in the same locations as the corresponding normal parts: two heads are always on the neck, extra fingers are always on the hand, extra toes on the foot. This is very difficult to explain on the basis of the theory that monsters come from the union of two foetuses or eggs, which was the explanation forced upon the preformationists by the nature of their views; but it was not at all difficult to explain on the basis of Maupertuis' concepts. He described the skeleton of a giant man, preserved in the Hall of Anatomy of the Academy in Berlin, with an extra vertebra in the lumbar region, inserted in a regular fashion between the ordinary vertebrae. How could this be the remains of a second foetus fused with the first? he asked.

But on Maupertuis' particulate theory, "chance, or the scarcity of family traits, will sometimes make rarer assemblages; and one will see born of black parents a white child, or perhaps even a black of white parents. . . ." ". . . there are elements so susceptible of arrangement, or in which recollection is so confused, that they become arranged with the greatest facility . . . ," elements which represent the condition in an ancestor rather than that in the immediate parent may enter into union in forming the embryo, producing resemblance to the ancestor rather than to the parent, but also "a total forgetfulness of the previous situation" may occur.

The Role of Mutations

Maupertuis thus came to the conclusion that hereditary variants are sudden, accidental products—mutations, to use the modern term. Moreover, since negroes could by mutation produce "whites" (i. e., albinos), it was clear that racial, or species, differences—the distinction was not too clear in the eighteenth century—are produced by mutations. To Maupertuis, exactly as to Hugo de Vries a century and a half later, a species was merely a mutant form that had become established in nature. The evidence for this was clear from the artificial breeds of domestic animals. As in the case of Charles Darwin a century later, it was in particular the pigeons that clinched the argument. "Nature contains the basis of all these variations: but chance or art brings them out. It is thus that those whose industry is applied to satisfying the taste of the curious are, so to say, creators of new species. We see appearing races of dogs, pigeons, canaries, which did not at all exist in Nature before. These were to begin with only fortuitous individuals; art and the repeated generations have made species of them. The famous Lyonnés [Lyonnet] every year created some new species, and destroyed that which was no longer in fashion. He corrects the forms and varies the colors: he has invented the species of the harlequin, the mopse, etc.". . .

If the ingenuity of man can produce species, why not nature, either by "fortuitous combinations of the particles of the seminal fluids, or effects of combining powers too potent or too weak among the particles" or by the action of the environment, such as the effect of climate or nutrition, on the hereditary particles. It is worth emphasizing, for it has been misunderstood, that Maupertuis raises the latter possibility only as one worthy of investigation; but clearly at this point he anticipated both Erasmus Darwin and [Jean-Baptiste de Monet de] Lamarck in suggesting the possibility of evolution through an inheritance of environmentally modified characters. Even so, it is the direct mutational action of heat or other factors on the hereditary material itself that Maupertuis seems most to have had in mind. "For the rest," he says, "although I suppose here that the basis of all these variations is to be found in the seminal fluids themselves, I do not exclude the influence that climate and foods might have. It seems that the heat of the torrid zone is more likely to foment the particles which render the skin black, than those which render it white: and I do not know to what point this influence of climate or of foods might extend, after long centuries of time."

Distribution of Species and Races

It is likewise clear that Maupertuis understood that most mutant forms are deleterious [abnormal or harmful] and at a disadvantage in comparison with the normal or wild types. "What is certain is that all the varieties which can characterize new species of animals and plants, tend to become extinguished: they are the deviations of Nature, in which she perseveres only through art or system. Her *works* always tend to resume the upper hand."

How, then, account for the distribution of different, races and species? The "thousands" of human varieties are insuperable difficulties for the preformationist; but by mutation, migration, and isolation they are readily accounted for by Maupertuis. Perhaps, he suggested, in the tropics all the peoples are dark of skin in spite of the interruptions caused by the sea, because of the heat of the torrid zone over a long period of time. The geographical isolation of Nature's deviations must play a part here, for "in travelling away from the equator, the color of the people grows lighter by shades. It is still very brown just outside the tropics; and one does not find complete whiteness until one has reached the temperate zone. It is at the limits of this zone that one finds the whitest peoples." Well, "men of excessive stature, and others of excessive littleness, are species of monsters; but monsters which can become peoples, were one to apply himself to multiplying them." Are there not races of giants and dwarfs? These "have become established, either by the suitability of climates, or rather because, in the time when they commenced to appear, they would have been chased into these regions by other men, who would have been afraid of the Colossi, or disdain the Pygmies.

"However many giants, however many dwarfs, however many blacks, may have been born among other men; pride or fear would have armed against them the greater part of mankind; and the more numerous species would have relegated these deformed races to the least habitable climates of the Earth. The Dwarfs will have retired toward the arctic pole: the Giants will have inhabited the Magellanic lands: the Blacks will have peopled the torrid zone." However naïve these anthropological conceptions may be—and they were an easy target for the sharp gibes of Voltaire—there is nevertheless a groping here for a truth that was only to be captured fully by Charles Darwin and Alfred Russel Wallace in a later day.

There is no naïvety, only pure genius, in these final words: "Could one not explain by that means [mutation] how from two individuals alone the multiplication of the most dissimilar species could have followed? They could have owed their first origination only to certain fortuitous productions, in which the elementary particles failed to retain the order they possessed in the father and mother animals; each degree of error would have produced a new species; and by reason of repeated deviations would have arrived at the infinite diversity of animals that we see today; which will perhaps still increase with tune, but to which perhaps the passage of centuries will bring only imperceptible increases."

Lamarck's Incorrect Theory of Evolution

L.J. JORDANOVA

One of the most important precursors of Charles Darwin's work on evolution was the theory proposed by French naturalist Jean-Baptiste de Monet de Lamarck, often referred to as transformism. Advanced and well-reasoned for its time, the theory ultimately proved to be incorrect. For one thing, as pointed out in this essay by L.J. Jordanova, a lecturer in history at the University of Essex, Lamarck accepted the doctrine of spontaneous generation. According to this view, living things could spring almost magically from nonliving things (as maggots appeared to spring from rotten meat before people realized that the maggots grew from eggs that flies laid in the meat). According to Jordanova, Lamarck also believed that physical changes that occurred in one generation could be inherited by the next. On the other hand, Lamarck did not, as some critics have charged, advocate teleology, the idea that nature has an underlying purpose.

L amarck arrived at his theory of the transformation of organic forms in 1799–1800 in the context of heated debates on extinction and fossils. In 1800, the opening lecture of his course at the Muséum revealed his new-found belief in the mutability of living nature. The undoubted novelty of his ideas and the controversies they provoked should not distract us from the equally important point that the roots of transformism went back to his earliest scientific work.

His theory rested on the following propositions: nothing in nature is constant; organic forms develop gradually from each other, and were not created all at once in their present form; all the natural sciences must recognize that nature has a history; and the laws governing living things have produced increasingly complex forms over immense periods of time.

Lamarck drew on numerous instances of transformations in both the inorganic and organic realms: the tides, chemical and geological mutations, spontaneous generation, processes of learning and development, ageing and adaptation. By 1800, the belief that 'nothing in nature is immutable' was a basic axiom in his natural philosophy. Despite their distinctive organic characteristics, the historical changes plants and animals underwent were only one aspect of the flux of nature.

Lamarck's Law of Nature

The most complete and best-known formulation of Lamarck's transformism was the *Zoological Philosophy* of 1809, which placed transformism in a broad biological context. The work must be seen as an ensemble, and an appreciation of its overall structure illuminates Lamarck's ideas of organic change. The *Zoological Philosophy* brought together classification, an analysis of the nature of life, especially in simple animals, and an account of the complex behavioural capacities of higher animals. . . .

His mind was, by 1809, firmly fixed on two projects: a natural history of invertebrates, and a study of man with particular reference to the nervous system and ethics. This twin focus, on the simplest and the most complex parts of the animal kingdom—central to transformism—was everywhere manifest in the *Zoological Philosophy*.

It was divided into three parts, each dealing with a distinct aspect of Lamarck's theories of living things: the natural history of animals, the physical causes of life, and the physical causes of what he called *sentiment*. Although this is best translated as 'feeling' or 'sentience', it is crucial to recognise that it was the biological capacity to receive sensations which concerned Lamarck, and not conscious acts.

The first part reassessed the classification of animals. From his botanical work, Lamarck was aware of the problem of imposing a system of classification on the natural world and then treating it as if it derived from nature itself. While other naturalists might have been content merely to acknowledge the artificiality of their systems, Lamarck was not. He strove to develop a way of coherently ordering animals while following nature's own plan. The project took on new significance with his transformism which offered a strong underpinning for a more natural classification because of its capacity to determine the actual order in which organisms were produced. For Lamarck, classificatory schemes ought to express the real relationships between living objects. Transformism and taxonomy were not mutually exclusive; although species were mutable they should still be named and their relationships with other forms specified. Before classifying, it helped if one had a theory to account for both the differences and the similarities between animals. Lamarck attributed the differences largely to the accidental effect of environmental factors. The similarities derived from 'the power of life', a law of nature which produced higher animals out of lower ones. Nature's use of basic templates to generate organic series of increasing complexity explained the observed relationships between forms.

The "Real" History of Animals

Before Lamarck, it had been customary to begin classificatory schemes with the most complex animals, gradually descending towards the more simple ones, as, indeed, he himself had done in early writings. Following the pattern he had established in the *System of Invertebrate Animals* (1800) and the *Researches on the Organization of Living Bodies* (1802), Lamarck devoted a lengthy chapter to the 'degradation and simplification in organisation from one end to the other of the animal chain, going from the most complex towards the most simple'. The concept of degradation was not a new one and would have been familiar to many of Lamarck's readers from [scientist George] Buffon's *Natural History*. Lamarck used it to suggest that, taken as a whole, the animal series displayed a striking gradation between complex and simple, from those with many faculties, a skeleton and vertebral column, to those entirely lacking these features. Having established the idea of a linkage between the main groups of animals, the 'principal masses' (Lamarck was careful to say that he was not concerned here with species), he reversed the direction of the series, making a chain of decreasing complexity into one of increasing complexity, starting with the most simple animals. The *real* history of the animal kingdom was conveyed by this new sequence. Natural history was now truly the history of nature.

Lamarck's ultimate goal was to understand the plan nature had followed and thereby to discover uniform, constant natural laws. He implied in early sections of the *Zoological Philosophy* that it was a law of nature to produce ever more complex living things which displayed regular, fine relationships between them. He was aware, of course, that the animal kingdom was not like that, and that the discrepancy was increasingly apparent the more one examined families and genera, rather than classes and 'principal masses'. Whereas nature's laws were responsible for the gradations among living things, the action of the environment accounted for specialised adaptations. . . .

For Lamarck the environment was a major part of natures; it operated according to natural laws, yet it was also in some sense the antithesis of life. 'Life' was the special power of nature to produce ever more elaborate, integrated and active organic beings. The inorganic, matter which made up the physical environment, left to itself, would decompose into its simplest constituents. Hence the natural world was composed of two forces constantly interacting in a dialectical manner; for an accurate classificatory system to be arrived at, they had to be unscrambled.

The influence of the environment was more evident in some cases than in others. Families of genera and species were groups with only fine gradations of organisation between members, there having been no extreme environmental changes to cause greater differences. Indeed, this seemed to act as a definition of 'family' as a taxonomic grouping. Had it acted without impediment, the 'power of life' would have produced a succession of regularly graduated forms starting with simple ones. Hence, if one looked at the general series of animals as a whole, the impact of the environment was clear in any deviation from this pattern.

Motion by Communication

Having argued in the first part of the book that species were not fixed, that animals could be arranged on a scale of increasing complexity which used human beings as the standard, and that classification should follow the chronological order of nature and hence be as 'natural' as possible, Lamarck set the scene for the analysis of the most important concept of the book, that of life.

The purpose of the second part of the book was to show that life was a purely physical phenomenon and to sketch out some of its basic properties. It therefore set out the first principles of biology with particular reference to zoology. The analysis of life was of considerable expiratory importance, for transformism rested on a number of presuppositions about the properties of the organic world, and 'the power of life' was itself a mechanism of transformation. Lamarck outlined his technique of finding out 'what life really is' by looking at simple animals with no special organs.

Between excited and communicated motion Lamarck drew a distinction the importance of which cannot be exaggerated: vital motion acted by excitation, motion in inert bodies was by communication. The definitions expressed his belief that although life could be analysed in physical terms, different physical principles should be invoked to explain inert matter. When motion was transmitted from one physical object to another, it was quite permissible to speak in terms of cause and effect for these could in fact be clearly separated. This was not the case in the living world, where cause and effect were inextricably intertwined as the nature and speed of the operations of the vertebrate nervous system illustrated. It should be emphasised that Lamarck did not thereby abdicate his responsibility as a scientist to find a physical explanation; he was merely asserting that the inorganic and organic worlds worked in different ways. Lamarck's sense that cause and effect could not be neatly separated in the organic realm was consistent with his emphasis on the dialectical relationship between organism and environment, and between different parts of the organism itself.

Identifying Degrees of Complexity

Locating the *source* of vital stimulation was Lamarck's next step. He thought that in simple animals it was the imponderable, invisible fluids in the environment, while in the most perfect animals, the excitatory cause of life was within each individual. Lamarck, following eighteenth-century medical and physiological traditions, located it principally in the nervous system. The environment as a source of vitality was exemplified by the spontaneous generation of rudimentary organisms. Lamarck stressed the role of fluids as mediators between organism and environment, as agents of all vital actions and as the mechanism whereby the number of organs and their associated functions increased. The superior vital energy of higher animals was manifested in their fluids, especially in nervous fluid, which acted on their passive parts (what Lamarck called 'containing' parts), the cellular tissue. All these remarks laid the foundations for the hierarchical distinctions

Lamarck wished to make, for example between animals and plants and between vertebrates and invertebrates. . . .

Lamarck was struck by the capacity of higher animals to change and adapt their behaviour through a highly complex nervous system. In human beings, it was the extraordinary capabilities of the brain and nervous system which characterised the species. Man—the masterpiece of nature—served as a vivid illustration of how the most intricate vital processes function. Lamarck therefore devoted the third part of the *Zoological Philosophy* to 'the physical causes of sentience', a section close to three hundred pages long in which, starting with first principles, he set out his approach to the analysis of nervous phenomena, including the operations of the human mind. His most important premises have already been mentioned: the rejection of innate ideas, the belief that all experience comes from the senses, and the assertion that structure and function are indissolubly linked. This part of Lamarck's masterpiece sometimes embarrassed subsequent generations because its emphasis on mental phenomena appeared to give weight to the commonly held view that Lamarck had illegitimately attributed consciousness and will to all animals, and hence had been guilty of psychologising biological phenomena. . . .

Lamarck's earliest work in botany had embodied the idea that identifying degrees of complexity was an essential step in developing an adequate classificatory system. This was achieved by establishing which were the most and the least complex forms, and then by filling in the space between them through assessing whole plants, not just isolated parts. He subsequently applied the method to zoology: 'if the lower end of this scale displays the minimum of animality, the other end necessarily displays the maximum'. The levels of complexity in plants were less striking to Lamarck than in animals; plant activity and life was relatively impoverished. The most vivid example of structural levels of complexity was provided by the animal nervous system. Lamarck did not say that one nervous system developed directly out of a previous one, but he pointed out that anatomical parts were added on in more complex animals, giving rise to more elaborate capacities. Anatomical, physiological and taxonomic levels were identical. . . .

Acquired Characteristics

When it came to mechanisms for transformism, Lamarck quite unselfconsciously assumed the inheritance of acquired characters, in the sense in which he understood it, to be so obvious and unexceptionable as to require very little comment. Since antiquity it had been believed that adaptations to changes which had taken place during the lifetime of an individual would be passed on to their offspring. Scientists continued to employ the idea, and the closely related one of habit, long after Lamarck's death. In addition to the inheritance of acquired characters, he simply postulated that nature was constantly in change and that life, by its very nature, became more complex with time. The environment certainly played a role as an agent of change. These ideas had been set out in his *Researches on the Organisation of Living Bodies* in 1802 and they remained the foundation of his later writings. It should be noted that for Lamarck transformism was a logical consequence of his views on the nature of the organic world; it was arrived at by deduction from the fundamental axioms governing all living things rather than by induction from a large number of empirical examples. This is not to say, of course, that Lamarck did not have a deep fund of natural historical knowledge on which to draw, for clearly he did. It was rather that since he regarded transformism as proven, he did not feel obliged to go and search for instances of it.

Adaptation was an empirical phenomenon which was important for Lamarck's arguments, and he understood it to be built into the nature of organisms. Biological need, grounded in the interaction of life and environmental forces, was a stimulus for action in animals without any 'effort' or 'will' being involved. The straightforward absence of the organ systems necessary for consciousness meant that most animals reacted to prevailing conditions by instinct rather than intelligence—a faculty which, according to Lamarck, nature had distributed with exceptional parsimony. The drive to adapt was so strong that animals responded automatically to stimuli from the outside world, and from inside their own bodies, like thirst or hunger. The emphasis on adaptation gave a teleological cast to Lamarck's arguments in that much stress was laid on the purposiveness of living things. His repeated personification further heightened the sense of nature having goals. Not only did it seem as if there was a pre-ordained purpose in nature, but Lamarck's language also implied that nature worked, even laboured, in the service of specific ends. This impression was an unfortunate product of Lamarck's rhetoric. In fact he thought there was no purpose outside nature, and human beings were merely one species among the many nature had produced. What commentators have construed as teleology, Lamarck saw as adaptation and progress generated by the interaction of physical forces.

The real achievement of the *Zoological Philosophy* was its fusion of a number of hitherto distinct areas—natural history, classification, physiology and psychology. It was the direct product of Lamarck's project for a treatise on biology, in that life and its unique characteristics were at its centre. At the same time, the mutability of organic forms was simply one example of uniform, natural laws which governed change in all bodies. The distinction between life and non-life notwithstanding, Lamarck saw nature as a single system of natural laws.

Erasmus Darwin's Version of Evolution

MICHAEL RUSE

One of the thinkers who advocated a theory of evolution be-
fore Charles Darwin did was his own grandfather, Erasmus
Darwin. The elder Darwin, a brilliant English physician, pub-
lished Zoonomia *in 1794–1796, in which he described the*
physical qualities of animals, as well as their diseases, but in
passing also set forth some crucial tenets of evolution (then
referred to as "transmutation"). This able summation of Eras-
mus Darwin's contributions to the development of modern
evolutionary theory is by science historian and philosopher
Michael Ruse, from his recent book The Evolution Wars.
Ruse makes the important point that Erasmus Darwin
thought that evolution was part of a grand plan designed by
God and therefore he was the first evolutionist to connect the
evolutionary and creationist arguments for life's origins.

The eighteenth century is called the Age of the Enlight-
enment, the time when the discoveries in science were
consolidated and extended and when in the arts and in literature people started to turn from the past and look to the
future. It is the time when we find such great writers and
critics as Voltaire; philosophers such as David Hume and
(a little later) Immanuel Kant; and the beginnings of social
science in the hands of such men as the Scottish political
economist, Adam Smith. Physics had had its great revolu-
tions in the two centuries previously. Chemistry was to

have its revolution toward the end of the century. . . . Biology was still looking forward. But the way was being prepared, thanks especially to the labors of two men. On the one hand, there was the Frenchman Georges Louis Leclerc, Comte de Buffon, author of the multivolumed *Histoire Naturelle* (from 1849 on), a discursive series of books that covered nature from one end to the next. Then on the other hand, there was the Swedish naturalist Carolus Linnaeus (Karl von Linné), whose ever-expanding *Systema Naturae* (first version 1735) introduced the modern system of organic classification, wherein every animal and plant can be fitted into its own unique place in the order of things.

Although neither was entirely successful in holding the dike, essentially both men had static pictures of nature. They had pictures that were, if not directly biblically based, then at least were views of life that might be called "Creationist," in the sense that God had created animals and plants basically in the forms that we see today, subject perhaps to a certain amount of variation, particularly of a degenerative kind. . . .

With the development of science . . . and the advances of literature and philosophy and political economy and more, people began to develop the confidence that not only is there change, but this can be permanent and brought about by us, through our own efforts. Moreover, whatever the naysayers may have claimed to the contrary, this was thought to be change for the good. Progressive change, in short. Such a philosophy, if one may so call it, was bound to have an effect on thinking about the organic world, and now as we shall see it truly did.

A Big Man in Every Sense

Erasmus Darwin, the grandfather of Charles, was a physician in the British Midlands in the second half of the eighteenth century. Famed for his skill—his diagnostic abilities were formidable—Darwin several times refused the earnest entreaties of poor oft-times mad King George the Third to come south and take on the role of court physician. He was happy in his station in life and particularly in his place in the country, which was just then experiencing the first wave of the Industrial Revolution. Around him enterprising engineers were putting to use the powers of coal and steam in the running of those machines that were to produce finished goods at a rate far more rapid than could ever be achieved by hand. The Midlands and the North of England were the sites of the action, and Darwin was in the thick of it, mixing with industrialists, scientists, engineers, and others, and himself contributing knowledge and advice drawn from his medical studies and experience, not to mention his general grasp of things scientific. A particular interest was the world of agriculture, something that had to experience no less of a revolution than industry, as people moved from the land to the cities, and as population numbers exploded, and hence as there was need to produce far more food with far less remaining available labor.

Erasmus Darwin was a man big in every sense of the word. His appetites were gargantuan. He loved his food so much that it was necessary to cut a semicircle in his table so that he could get close to the action. Preparing for one of his visits required considerable forethought and expense. Expensive dishes—preferably many of them—were expected and appreciated. But Darwin gave as he received. He was a wonderful conversationalist and a much-loved friend, valued for his sensible advice. . . .

One of Darwin's closest friends was the potter Josiah Wedgwood, he who was responsible for the development of the British china trade—cups and saucers, plates and dishes, as well as vases and other objects of great beauty. In this prerailway age, the chief mode of transportation—especially safe and careful transportation—was by water. Supplementing the sea and the rivers, the eighteenth century was a time of great canal building: something that required an intimate knowledge of geology, especially when there were questions of boring tunnels through mountains.

Wedgwood was a major figure in this work, and Darwin was in the midst of this activity, looking and searching and thinking and exclaiming. "I have lately travel'd two days journey into the bowels of the earth, with three most able philosophers, and have seen the Goddess of Minerals naked, as she lay in her inmost bowers."

Erasmus Darwin was absolutely fascinated by discoveries such as these. Looking back two centuries later, there is no "smoking gun" that proves definitively just what it was that tipped him toward evolutionism or (as, in those days, he would have called it) transmutationism. Most probably it was the marine remains (shells and fossil fish) found hidden away in that mountain, in the middle of England, where he journeyed with his companions. Certainly, soon thereafter Darwin adopted *E conchis omnia* (Everything from shells) as his personal motto, and to celebrate he had the phrase painted on the door of his own carriage. He did not rush into print, however. Setting a pattern that was to be followed by his grandson Charles, Erasmus Darwin took some 20 years before he felt ready to announce his thinking to the outside world.

The Ideas in *Zoonomia*

His ideas were first written about explicitly in his major medical treatise *Zoonomia*, although one could hardly say that the treatment there was particularly systematic. Darwin made little or no attempt to disentangle the various threads of his thinking. Claims about the *fact* of evolution were mingled with ideas about the *paths* of evolution, and then threaded through the whole discussion were all sorts of hypotheses and speculation about the *causes* of evolution. Quite often he would start a paragraph talking about paths and then end up talking about causes. Or he would start off talking about causes and end up arguing for the general fact. He may have been an innovative thinker; he was no great systematist.

Trying our best to disentangle his thinking, we find that

probably there were two direct arguments that Erasmus Darwin put forward for the fact of evolution. First of all, he was much impressed by the analogy that he presumed between individual development and group development. If we can transform the individual—"from the feminine boy to the bearded man, and from the infant girl to the lactescent woman"—then why should we not transmute the group? Second, he thought very significant the similarities that he saw holding between the parts of the members of quite different species. These similarities, which today we call "homologies," were taken—as, indeed, they are taken today—to be evidence of common ancestry. Although, as I have just said, it is almost certain that it was fossil discoveries that made Darwin an evolutionist in the first place, he did not really bring in the fossils as a major piece of information in favor of the fact of evolution. They are mentioned but not as an important plank in the evidential foundation.

Today, we would surely want to use the fossils as evidence of pathways. Erasmus Darwin made no move in this direction either, although in fairness he had virtually none of the evidence that today makes the fossil record so important a source of information. As we shall see in a moment, he had an overall vision of the path of evolution, but as far as the specifics are concerned, he said little. In his opinion, the best source of information for actual pathways lay in the natures of living organisms. Take the presumed transition from sea to land. Erasmus Darwin touched on the peculiarities of animals like whales, seals, and frogs. He seemed to think that animals of this kind are somehow representative of those transitional forms that must have existed when life made its move from the sea to the land. Since we have such hybrid types today, it is reasonable to assume that they existed in the past, and these types today give us some clue as to their former nature.

What interested Erasmus Darwin more was the question of causes. He collected and offered all sorts of jumbled anecdotal bits and pieces of information. As you might expect, given that Darwin was living in a particularly important agricultural part of England, many of his suggestions were based on the folklore of animal and plant breeders. Indeed, Darwin spoke explicitly of "the great changes introduced into various animals by artificial and accidental cultivation." He was a strong supporter of the idea that characteristics acquired by an organism in one generation can be passed straight to members of the next generation. He instanced the docking of dogs' tails, Darwin believed that this practice eventually results in the birth of animals without tails at all, and therefore without any need of docking. This inheritance of acquired characteristics is today known as "Lamarckism" after the great French evolutionist of that name, although it should be noted that Lamarck's writings came at least a decade after Darwin put pen to paper. . . .

Naturally, as a physician, Erasmus Darwin was much interested in the nature of the mind and in the ways in which mental attributes can affect and be affected by physical causes. The popular psychological theory of his day—the brainchild of the eighteenth-century thinker David Hartley—was known as "associationism." In line with the general associationist position, Erasmus Darwin thought that habits and experiences could lead to new beliefs, and that these beliefs could be passed straight on thanks to reproduction. Hence, people's mental attributes could be a result of things having happened in the past to members of earlier generations. . . . Also, most interestingly, there was an anticipation of an idea that was promoted by grandson Charles. Erasmus Darwin thought that it was entirely possible that the body throws off small parts; these are carried around, presumably by the blood; and finally they are gathered in and transmitted via the sex organs. This supposedly gave a physiological backing to the already mentioned Lamarckism. The blacksmith's arms get stronger and stronger through use. These newly developed arms cast off modified particles that go down to the sex organs. And so the children of the blacksmith are born with strong arms as part of their biological heritage. . . .

Obsessed by the Concept of Progress

Fortunately, the overall ideas of Erasmus Darwin are not too hard to follow. We start at the bottom with the most primitive form, what was then often called the "monad," and we work our way up to the most complex and best form, what was then (unself-consciously) known as the "man." From butterfly (monarch) to king (monarch), as he expressed himself on another occasion. From that which is totally without value to that which we value above all else. A progressive rise up the chain of life. Yet, straight-forward though this vision may be, I do want to make a couple of points before we move on.

The first is a general point but applied specifically to Erasmus Darwin. It is about the way in which we should treat figures in the past: There is a temptation to go too far in one way or the other, to see too many virtues or too many faults. Either we see the historical figure as a pure genius, with no flaws, and as having anticipated just about everything. Or we see him or her as a real fool, who found his or her way in the history books by chance or default or even fraud. Erasmus Darwin is a good case in point. On the one hand, he surely did come up with evolution as fact long before a lot of other people. He was right on there. Moreover, he did pick up on some good points. Fossils are important. The similarities between the bone structures of very different organisms are puzzling at the least, and surely suggestive of some hidden links. And embryology? Well, we do develop from primitive beginnings, so why should not the same be true of life itself?

On the other hand, if ever anyone was credulously open to absurd arguments it was Erasmus Darwin. There was no systematic treatment with things properly quantified—the very things that, by the end of the eighteenth century, one took for granted in the physical sciences. Again and again the reader would get something far more suited for Ripley's *Believe It or Not* than for anything with pretensions to being serious science. . . .

What am I trying to tell you? Basically, that at a certain level there was something rather ambiguous or questionable about both the quality and the status of the evolutionary speculations of Dr. Erasmus Darwin. Of course, we today would think this; but the point I am making is that even in the eyes of his contemporaries the ideas of Darwin were somewhat dubious or suspect. Which raises another question. If Darwin was indeed writing at such a loose or unsubstantiated level, why was he driven to do so? He was no fool, nor was he an unsophisticated thinker about technical issues. I told you that he truly had a great and justified reputation as a physician. Why then did he write as he did about evolution, and why was it that others at the time responded favorably to his ideas?

The answer has been given already. Darwin and his followers were absolutely obsessed with the new philosophy of the day: the philosophy or ideology of progress. For Darwin and his supporters, the Industrial Revolution—which was now going ahead at full steam, to use an apt metaphor—was the best thing that had ever happened to rural, sleepy, church-dominated England. What was needed, therefore, was a complete change of worldview. A worldview making central the success of machines and of the men of purpose who devised and drove them. That is to say, a worldview making central the achievements and aims of Darwin himself and of his industrialist friends. Evolution for Darwin, and for his supporters, was very much part and parcel of this philosophy or vision. . . .

All in all, therefore, the evolutionism of Dr. Darwin was the industrialist's philosophy of action made flesh—or embedded in the rocks! One goes from "an embryon point, or microscopic ens!" to "imperious man, who rules the bestial crowd." At work here is a full-blown circular argument, or perhaps more charitably one might say a feedback argument. You start with the idea of progress, the philosophy of the British industrialist. You read this into nature. And then you read it right back to confirm your philosophy. "All nature exists in a state of perpetual improvement . . . the world may still be said to be in its infancy, and continue to improve FOR EVER and EVER."

Belief in an Unmoved Mover

Is this the philosophy of a man who has turned his back against religion? In a sense, this has to be true. Erasmus Darwin was certainly putting himself in opposition to conventional Christianity. For the Christian, the overall history of the world is one of miraculous creation, of subsequent sin and fall, and of the need for redemption that comes through, and only through, God's grace. Christ's great sacrifice on the cross and his miraculous rising from the dead wash away the sins of us all. For the Christian, therefore, Providence is the key to understanding history and the future. We humans can do nothing, save only with God's help and love. Darwin, as a progressionist, was arguing strongly that we humans are capable of improving our lot ourselves. So, in this sense, quite apart from the fact that as an evolutionist he had no place for the creation story of Genesis, Darwin was putting himself against traditional religion.

However, one should not at once conclude that Darwin was an atheist, or even an agnostic in the sense of having any doubts about God's existence. Darwin was no Christian, but like many intellectuals of his age (including many of the early American presidents), Darwin believed in a God who was an unmoved mover. He believed in a God who has put things in motion and who then stands back and watches how things work out through the agency of unbroken law. To use the technical language of scholars, Darwin was a deist, as opposed to a theist, traditionally a Christian, a Jew, and a Muslim. A deist sees the greatest mark of God's power and forethought in the working out of unbroken law, as opposed to the theist who sees God's power in direct intervention, that is, in miracles.

Using a modern metaphor, what one might say is that Darwin's god—the god of the deist—has preprogrammed the world so that he did not have to intervene further. Evolution, therefore, can be seen as the greatest triumph of God. It is the strongest proof of his existence. It is certainly not something that disproves the need for or existence of a Creator or Designer. In Darwin's own words, "What a magnificent idea of the infinite power of *The Great Architect! The Cause of Causes! Parent of Parents! Ens Entium!*"

I am not now sure that you would want to say that Darwin's evolutionism was a religious theory, nor even am I quite sure what that might mean. But this is the first moment at which you should start to realize that the science-religion relationship—the relationship in the context of evolution—is more complex than you might have thought. Those people (and there are many) who seem to think that evolutionists become atheists in the morning and then think up their theories in the afternoon, as a kind of bad joke, could not be more mistaken. Certainly, Erasmus Darwin—the man who can first claim unambiguously the label of "evolutionist"—became an evolutionist as much because of his religious beliefs as despite them. And that is a good point on which to move forward.

The Biblical View of Life's Origins Remains Dominant

RICHARD DAWKINS

None of the evolutionary theories proposed from ancient times made much impact on how most people viewed the creation of earth and humans. The majority, including most scientists, simply accepted the traditional biblical view, that God created plants and animals miraculously, so there was not yet a major controversy. This essay, by Richard Dawkins, a professor of zoology at Oxford University, discusses that traditional view, which remained the dominant one until Charles Darwin made a stir with his Origin of Species *in the mid-1800s. As Dawkins explains, the traditional view was summarized in the famous analogy of a watch and its maker, the so-called "argument from design." Dawkins cites the example of eighteenth-century theologian William Paley, the first writer to make the watchmaker analogy. Before Darwin's time, the vast majority of people accepted without question Paley's thesis of divine creation.*

We animals are the most complicated things in the known universe. The universe that we know, of course, is a tiny fragment of the actual universe. There may be yet more complicated objects than us on other planets, and some of them may already know about us. But this doesn't alter the point that I want to make. Complicated things, everywhere, deserve a very special kind of explanation. We want to know how they came into existence and why they are so complicated. The explanation, as I shall argue, is likely to be broadly the same for complicated things everywhere in the universe; the same for us, for chimpanzees, worms, oak trees and monsters from outer space. On the other hand, it will not be the same for what I shall call 'simple' things, such as rocks, clouds, rivers, galaxies and quarks. These are the stuff of physics. Chimps and dogs and bats and cockroaches and people and worms and dandelions and bacteria and galactic aliens are the stuff of biology.

The difference is one of complexity of design. Biology is the study of complicated things that give the appearance of having been designed for a purpose. Physics is the study of simple things that do not tempt us to invoke design. At first sight, man-made artefacts like computers and cars will seem to provide exceptions. They are complicated and obviously designed for a purpose, yet they are not alive, and they are made of metal and plastic rather than of flesh and blood. . . .

I said that physics is the study of simple things. . . . Physics appears to be a complicated subject, because the ideas of physics are difficult for us to understand. Our brains were designed to understand hunting and gathering, mating and child-rearing: a world of medium-sized objects moving in three dimensions at moderate speeds. . . . We think that physics is complicated because it is hard for us to understand, and because physics books are full of difficult mathematics. . . .

Physics *books* may be complicated, but physics books, like cars and computers, are the product of biological objects—human brains. The objects and phenomena that a physics book describes are simpler than a single cell in the body of its author. And the author consists of trillions of those cells, many of them different from each other, organized with intricate architecture and precision-engineering into a working machine capable of writing a book. . . . Our brains are no better equipped to handle extremes of complexity than extremes of size and the other difficult extremes of physics. Nobody has yet invented the mathematics for describing the total structure and behaviour of such an object as a physicist, or even of one of his cells. What we can do is understand some of the general principles of how living things work, and why they exist at all.

Each of Us Is a Machine

This was where we came in. We wanted to know why we, and all other complicated things, exist. And we can now answer that question in general terms, even without being able to comprehend the details of the complexity itself. To take an analogy, most of us don't understand in detail how an airliner works. Probably its builders don't comprehend it fully either: engine specialists don't in detail understand wings, and wing specialists understand engines only vaguely. Wing specialists don't even understand wings with full mathematical precision: they can predict how a wing will behave in turbulent conditions, only by examining a model in a wind tunnel or a computer simulation—the sort of thing a biologist might do to understand an animal. But however incompletely we understand how an airliner works, we all understand by what general process it came into existence. It was designed by humans on drawing boards. Then other humans made the bits from the drawings, then lots more humans (with the aid of other machines designed by humans) screwed, rivetted, welded or glued the bits together, each in its right place. The process by which an airliner came into existence is not fundamentally mysterious to us, because humans built it. The systematic putting together of parts to a purposeful design is something we know and understand, for we have experienced it at first hand, even if only with our childhood Meccano or Erector set.

What about our own bodies? Each one of us is a machine, like an airliner only much more complicated. Were we designed on a drawing board too, and were our parts assembled by a skilled engineer? The answer is no. It is a surprising answer, and we have known and understood it for only a century or so. When Charles Darwin first explained the matter, many people either wouldn't or couldn't grasp it. I myself flatly refused to believe Darwin's theory when I first heard about it as a child. Almost everybody throughout history, up to the second half of the nineteenth century, has firmly believed in the opposite—the Conscious Designer theory. Many people still do, perhaps because the true, Darwinian explanation of our own existence is still, remarkably, not a routine part of the curriculum of a general education. It is certainly very widely misunderstood.

A Stone vs. a Watch

The watchmaker [is an image] . . . from a famous treatise by the eighteenth-century theologian William Paley. His *Natural Theology—or Evidences of the Existence and Attributes of the Deity Collected from the Appearances of Nature*, published in 1802, is the best-known exposition of the 'Argument from Design', always the most influential of the arguments for the existence of a God. It is a book that I greatly admire, for in his own time its author succeeded in doing what I am struggling to do now. He had a point to make, he passionately believed in it, and he spared no effort to ram it home clearly. He had a proper reverence for the complexity of the living world, and he saw that it demands a very special kind of explanation. The only thing he got wrong—admittedly quite a big thing!—was the explanation itself. He gave the traditional religious answer to the riddle, but he articulated it more clearly and convincingly than anybody had before. The true explanation is utterly different, and it had to wait for one of the most revolutionary thinkers of all time, Charles Darwin.

Paley begins *Natural Theology* with a famous passage:

In crossing a heath, suppose I pitched my foot against a *stone*, and were asked how the stone came to be there; I might possibly answer, that, for anything I knew to the contrary, it had lain there for ever: nor would it perhaps be very easy to show the absurdity of this answer. But suppose I had found a *watch* upon the ground, and it should be inquired how the watch happened to be in that place; I should hardly think of the answer which I had before given, that for anything I knew, the watch might have always been there.

Paley here appreciates the difference between natural physical objects like stones, and designed and manufactured objects like watches. He goes on to expound the precision with which the cogs and springs of a watch are fashioned, and the intricacy with which they are put together. If we found an object such as a watch upon a heath, even if we didn't know how it had come into existence, its own precision and intricacy of design would force us to conclude

> that the watch must have had a maker: that there must have existed, at some time, and at some place or other, an artificer or artificers, who formed it for the purpose which we find it actually to answer; who comprehended its construction, and designed its use.

Nobody could reasonably dissent from this conclusion, Paley insists, yet that is just what the atheist, in effect, does when he contemplates the works of nature, for:

> every indication of contrivance, every manifestation of design, which existed in the watch, exists in the works of nature; with the difference, on the side of nature, of being greater or more, and that in a degree which exceeds all computation.

Paley drives his point home with beautiful and reverent descriptions of the dissected machinery of life, beginning with the human eye, a favourite example which Darwin was later to use. . . . Paley compares the eye with a designed instrument such as a telescope, and concludes that 'there is precisely the same proof that the eye was made for vision, as there is that the telescope was made for assisting it'. The eye must have had a designer, just as the telescope had.

No Purpose in Mind

Paley's argument is made with passionate sincerity and is informed by the best biological scholarship of his day, but it is wrong, gloriously and utterly wrong. The analogy between telescope and eye, between watch and living organism, is false. All appearances to the contrary, the only watchmaker in nature is the blind forces of physics, albeit deployed in a very special way. A true watchmaker has foresight: he designs his cogs and springs, and plans their interconnections, with a future purpose in his mind's eye. Natural selection, the blind, unconscious, automatic process which Darwin discovered, and which we now know is the explanation for the existence and apparently purposeful form of all life, has no purpose in mind. It has no mind and no mind's eye. It does not plan for the future. It has no vision, no foresight, no sight at all. If it can be said to play the role of watchmaker in nature, it is the *blind* watchmaker.

A Theologian Dates the Creation

JAMES USSHER

In 1650, a Protestant bishop named James Ussher laid a firm foundation that later fundamentalist Christians would use to support their arguments against the theory of evolution. Ussher carefully traced the lives of the kings, prophets, judges, and other major figures mentioned in the Old Testament. In his Annals of the Old Testament from the Beginning of the World, *he listed their life spans and the dates he contended they had lived, traced them backward, and concluded that the creation had occurred in 4004 B.C. What follows is the opening section of Ussher's long and convoluted list. From this evidence, fundamentalist Christians conclude that the world is not old enough for evolution to have taken place. (Among the abbreviations he uses are AM, for "Year of the World from the Creation," and JP, for the date according to the Julian calendar.)*

The First Age of the World

1a AM, 710 JP, 4004 BC

1. In the beginning God created the heaven and the earth. Ge 1:1 This beginning of time, according to our chronology, happened at the start of the evening preceding the 23rd day of October in the year of the Julian calendar, 710.

2. On the first day Ge 1:1–5 of the world, on Sunday, October 23rd, God created the highest heaven and the angels. When he finished, as it were, the roof of this building, he started with the foundation of this wonderful fabric of the

---

James Ussher, *The Annals of the Old Testament from the Beginning of the World*, 1650.

world. He fashioned this lower most globe, consisting of the deep and of the earth. Therefore all the choir of angels sang together and magnified his name. Job 38:7 When the earth was without form and void and darkness covered the face of the deep, God created light on the very middle of the first day. God divided this from the darkness and called the one "day" and the other "night".

3. On the second day Ge 1:6–8 (Monday, October 24th) after the firmament or heaven was finished, the waters above were separated from the waters here below enclosing the earth.

4. On the third day Ge 1:9–13 (Tuesday, October 25th) when these waters below ran together into one place, the dry land appeared. From this collection of the waters God made a sea, sending out from here the rivers, which were to return there again. Ec 1:7 He caused the earth to bud and bring forth all kinds of herbs and plants with seeds and fruits. Most importantly, he enriched the garden of Eden with plants, for among them grew the tree of life and the tree of knowledge of good and evil. Ge 2:8,9

5. On the fourth day (Wednesday, October 26th) the sun, the moon and the rest of the stars were created.

6. On the fifth day (Thursday, October 27th) fish and flying birds were created and commanded to multiply and fill the sea and the earth.

7. On the sixth day (Friday, October 28th) the living creatures of the earth were created as well as the creeping creatures. Last of all, man was created after the image of God, which consisted principally in the divine knowledge of the mind, Col 3:10 in the natural and proper sanctity of his will. Eph 4:24 When all living creatures by the divine power were brought before him, Adam gave them their names. Among all of these, he found no one to help him like himself. Lest he should be destitute of a suitable companion, God took a rib out of his side while he slept and fashioned it into a woman. He gave her to him for a wife, establishing by it the law of marriage between them. He blessed them and bade

them to be fruitful and multiply. God gave them dominion over all living creatures. God provided a large portion of food and sustenance for them to live on. To conclude, because sin had not yet entered into the world, God saw every thing that he had made, and, behold, it was very good. And the evening and the morning were the sixth day. Ge 1:31

8. Now on the seventh day, (Saturday, October 29th) when God had finished his work which he intended, he then rested from all labour. He blessed the seventh day and ordained and consecrated the sabbath Ge 2:2,3 because he rested on it Ex 31:17 and refreshed himself. Nor as yet (for ought appears) had sin entered into the world. Nor was there any punishment given by God, either upon mankind, or upon angels. Hence it was, that this day was set forth for a sign, as well as for our sanctification in this world Ex 31:13 of that eternal sabbath, to be enjoyed in the world to come. In it we expect a full deliverance from sin and its dregs and all its punishments. Heb 4:4,9,10

9. After the first week of the world ended, it seems that God brought the newly married couple into the garden of Eden. He charged them not to eat of the tree of knowledge of good and evil but left them free to eat of everything else.

10. The Devil envied God's honour and man's obedience. He tempted the woman to sin by the serpent. By this he got the name and title of the old serpent. Re 12:9 20:2 The woman was beguiled by the serpent and the man seduced by the woman. They broke the command of God concerning the forbidden fruit. Accordingly when sought for by God and convicted of this crime, each had their punishments imposed on them. This promise was also given that the seed of the woman should one day break the serpent's head. Christ, in the fulness of time should undo the works of the Devil. 1Jo 3:8 Ro 16:20 Adam first called her Eve because she was then ordained to be the mother, not only of all that should live this natural life, but, of those also who should live by faith in her seed. This was the promised Messiah as Sarah also later was called the mother of the faithful. 1Pe 3:6 Ga 4:31

11. After this our first parents were clothed by God with raiment of skins. They were expelled from Eden and a fiery flaming sword set to keep the way leading to the tree of life so that they should never eat of that fruit which they had not yet touched. Ge 3:21,22 It is very probable, that Adam was turned out of paradise the same day that he was brought into it. This seems to have been on the 10th day of the world. (November 1st) On this day also, in remembrance of so remarkable an event the day of atonement was appointed Le 23:27, and the yearly fast, spoken of by Paul, Ac 27:9. . . . On this feast all, strangers as well as native Israelites, were commanded to afflict their souls that every soul which should not afflict itself upon that day should be destroyed from among his people, Le 16:29 23:29

12. After the fall of Adam, Cain was the first of all mortal men that was born of a woman. Ge 4:1

130d AM, 840 JP, 3874 BC

13. When Cain, the firstborn of all mankind, murdered Abel, God gave Eve another son called Seth. Ge 4:25 Adam had now lived 130 years. Ge 5:3 From whence it is gathered, that between the death of Abel and the birth of Seth, there was no other son born to Eve. For then, he should have been recorded to have been given her instead of him. Since man had been on the earth 128 years and Adam and Eve had other sons and daughters Ge 5:4 the number of people on the earth at the time of this murder could have been as many as 500,000. Cain might justly fear, through the conscience of his crime, that every man that met him would also slay him. Ge 4:14,15

235d AM, 945 JP, 3769 BC

14. When Seth was 105 years old, he had his son, Enos. This indicates the lamentable condition of all mankind. For even then was the worship of God wretchedly corrupted by the race of Cain. Hence it came, that men were even then so distinguished, that they who persisted in the true

worship of God, were known by the name of the children of God. They who forsook him, were termed the children of men. Ge 4:26 6:1,2

325d AM, 1035 JP, 3679 BC
15. Cainan, the son of Enos was born when his father was 90 years old. Ge 5:10

395d AM, 1105 JP, 3609 BC
16. Mahalaleel was born when Cainan his father was 70 years old. Ge 5:12

460d AM, 1170 JP, 3544 BC
17. Jared was born when his father Mahalaleel was 65 years old. Ge 5:15

622d AM, 1332 JP, 3382 BC
18. Enoch was born when his father Jared was 162 years old. Ge 5:18

687d AM, 1397 JP, 3317 BC
19. Methuselah was born when Enoch his father was 65 years old. Ge 5:25

874d AM, 1584 JP, 3130 BC
20. Lamech was born when his father Methuselah was 187 years old. Ge 5:25

930d AM, 1640 JP, 3074 BC
21. Adam, the first father of all mankind, died at the age of 930 years. Ge 5:5

987d AM, 1697 JP, 3017 BC
22. Enoch, the 7th from Adam at the age of 365 years, was translated by God in an instant, while he was walking with him that he should not see death. Ge 5:23,24 Heb 11:5

1042d AM, 1752 JP, 2962 BC
23. Seth, the son of Adam died when he was 912 years old. Ge 5:8

1056d AM, 1766 JP, 2948 BC

24. Noah, the 10th from Adam, was born when his father Lamech was 182 years old. Ge 5:29

1140d AM, 1850 JP, 2864 BC

25. Enos, the 3rd from Adam, died when he was 905 years old. Ge 5:11

1235d AM, 1945 JP, 2769 BC

26. Cainan, the 4th from Adam, died when he was 910 years old. Ge 5:14

1290d AM, 2000 JP, 2714 BC

27. Mahalaleel, the 5th from Adam, died when he was 892 years old. Ge 5:17

1422d AM, 2132 JP, 2582 BC

28. Jared, the 6th from Adam, died when he was 962 years old. Ge 5:20

1536a AM, 2245 JP, 2469 BC

29. Before the deluge of waters upon the whole wicked world, God sent Noah, a preacher of righteousness to them, giving them 120 years to repent of their evil ways. 1Pe 3:20 2Pe 2:5 Ge 6:3

1556d AM, 2266 JP, 2448 BC

30. Noah was 500 years old when his 1st son, Japheth was born. Ge 5:32 10:21

1558d AM, 2268 JP, 2446 BC

31. Noah's 2nd son, Shem, was born 2 years later because 2 years after the flood, Shem was 100 years old. Ge 11:10

1651d AM, 2361 JP, 2353 BC

32. Lamech, the 9th from Adam, died when he was 777 years old. Ge 5:31

1656a AM, 2365 JP, 2349 BC

33. Methuselah, the 8th from Adam, died when he was

969 years old. He lived the longest of all men yet died before his father. Ge 5:27,24

34. Now in the 10th day of the second month of this year (Sunday, November 30th) God commanded Noah that in that week he should prepare to enter into the Ark. Meanwhile the world, totally devoid of all fear, sat eating and drinking and marrying and giving in marriage. Ge 7:1,4,10 Mt 24:38

35. In the 600th year of the life of Noah, on the 17th day of the second month, (Sunday, December 7th), he with his children and living creatures of all kinds had entered into the Ark. God sent a rain on the earth 40 days and 40 nights. The waters continued upon the earth 150 days, Ge 7:4,6,11–13,17,24

36. The waters abated until the 17th day of the 7th month, (Wednesday, May 6th) when the Ark came to rest upon one of the mountains of Ararat. Ge 8:3,4

37. The waters continued receding until on the 1st day of the 10th month (Sunday, July 19th) the tops of the mountains were seen. Ge 8:5

38. After 40 days, that is on the 11th day of the 11th month (Friday, August 28th) Noah opened the window of the Ark and sent forth a raven. Ge 8:6,7

39. 7 days later, on the 18th day of the 11th month (Friday, September 4th) as may be deduced from the other 7 days mentioned in Ge 8:10, Noah sent out a dove. She returned after 7 days. 25th day of the 11th month, (Friday, September 11th) he sent her out again and about the evening she returned bringing the leaf of an olive tree in her bill. After waiting 7 days more, 2nd day of the 12th month, (Friday, September 18th) he sent the same dove out again, which never returned. Ge 8:8,12

The Second Age of the World
1657a AM, 2366 JP, 2348 BC

40. When Noah was 601 years old, on the 1st day of the 1st month (Friday, October 23rd), the 1st day of the new

post-flood world, the surface of the earth was now all dry. Noah took off the covering of the Ark. Ge 8:13

41. On the 27th day of the 2nd month (Thursday, December 18th), the earth was entirely dry. By the command of God, Noah went forth with all that were with him in the Ark. Ge 8:14,15,19

42. When he left the Ark, Noah offered to God sacrifices for his blessed preservation. God restored the nature of things destroyed by the flood. He permitted men to eat flesh for their food and gave the rainbow for a sign of the covenant which he then made with man. Ge. 8:15-9:17

THE
HISTORY
OF
ISSUES

CHAPTER 2

Evolution and Natural Selection Create a Controversy

Chapter Preface

I t is almost certain that the great controversy over evolution that erupted in the mid-nineteenth century when Charles Darwin published his *Origin of Species* would not have occurred if Darwin had not sailed on the HMS *Beagle* in 1831. Indeed, had he not studied various plant and animal species in South America and in the Gálapagos Islands during the trip, it is unlikely that *Origin* would ever have been written. Charles Darwin might well have become a clergyman, the profession his father had chosen for him. In that case, Alfred Russel Wallace, who devised a theory of natural selection in the same period Darwin did, would have published the seminal version of that theory. The difference is that Wallace did not have the mass of data Darwin had collected during and after the *Beagle* journey; so Wallace's book probably would not have made as much of a stir as Darwin's book did.

A friend alerted Darwin to the voyage that would end up changing both his life and the course of science. The *Beagle* was preparing to sail, the friend said, and its captain, Robert FitzRoy, was looking for a naturalist to go along and document the expedition's discoveries. The eager young Darwin, who had long dreamed of being a naturalist, got the job.

The exciting journey took Darwin first to Argentina, in South America, where the ship sailed down rivers and explored some of the country's inland regions. Darwin observed Argentine ostriches, which, he learned, existed in two separate varieties. He wondered why God would create two similar yet plainly distinct species of ostrich and place them in the same geographic space. He did not mean to question God's logic. But the young man's instincts told him that some powerful, unknown natural principle or force was at work here.

This notion was strongly reinforced after the *Beagle* reached the distant and mysterious Gálapagos chain in the southern Pacific. Darwin found the primitive islands inhabited by gigantic tortoises, huge lizards, bright red crabs, and all manner of exotic birds. From one island to the next, he observed, the giant tortoises and their shells varied considerably in shape. As in the case of the South American ostriches, he wondered why more than one kind of giant tortoise existed in the same region. Even more striking to Darwin were the variations he observed in many of the birds of the Gálapagos. A single bird species often displayed noticeable differences from island to island.

At the time, Darwin did not immediately conclude that these physical differences in the local animals were evidence of evolution. Yet in the weeks, months, and years that followed, such an explanation began to materialize in his mind; at first it was tentative and largely unsupported, but as time went on it became increasingly compelling in its powerful logic, beauty, and simplicity. He began to see that the Gálapagos Islands had once been completely barren of life. Over the course of thousands of years, finches from the mainland occasionally got caught in wind currents and ended up on the islands. Finding an environment very different from that of the mainland, the birds had to find different kinds of foods and adjust to different habits. A tiny bit at a time, as thousands of generations of finches lived and died, the species underwent physical and behavioral changes in an effort to adjust to its new environment.

Darwin began to apply this same logic to other animals in the Gálapagos and elsewhere. That steadily channeled his thinking in the direction of evolution and, eventually, the doctrine of natural selection, in which nature selects the strongest and most adaptable species to survive. For these reasons, the *Beagle*'s round-the-world journey proved to be the seminal experience of Darwin's career and one of the great milestones in the history of science.

Darwin Discovers Natural Selection

DANIEL C. DENNETT

Since many people had conceived the notion of evolution be-
fore Darwin did, his seminal idea was not evolution itself, but
rather a detailed and believable explanation of the physical
processes driving evolution. The most important factor in
these processes, he said, is the doctrine of natural selection,
which is better known as "survival of the fittest." As shown in
this clearly written tract by Daniel C. Dennett, director of the
Center for Cognitive Studies at Tufts University, Darwin got
the inspiration for the idea from reading the renowned 1798
essay on population penned by Englishman Thomas Malthus.

Darwin's project in *Origin* can be divided in two: to
prove *that* modern species were revised descendants
of earlier species—species had evolved—and to show *how*
this process of "descent with modification" had occurred.
If Darwin hadn't had a vision of a mechanism, natural selection, by which this well-nigh-inconceivable historical
transformation could have been accomplished, he would
probably not have had the motivation to assemble all the
circumstantial evidence that it had actually occurred. Today we can readily enough imagine proving Darwin's first
case—the brute historic fact of descent with modification—quite independently of any consideration of natural
selection or indeed any other mechanism for bringing
these brute events about, but for Darwin the idea of the
mechanism was both the hunting license he needed, and

an unwavering guide to the right questions to ask.

The idea of natural selection was not itself a miraculously novel creation of Darwin's but, rather, the offspring of earlier ideas that had been vigorously discussed for years and even generations. Chief among these parent ideas was an insight Darwin gained from reflection on the 1798 *Essay on the Principle of Population* by Thomas Malthus, which argued that population explosion and famine were inevitable, given the excess fertility of human beings, unless drastic measures were taken. The grim Malthusian vision of the social and political forces that could act to check human overpopulation may have strongly flavored Darwin's thinking (and undoubtedly has flavored the shallow political attacks of many an anti-Darwinian), but the idea Darwin needed from Malthus is purely logical. It has nothing at all to do with political ideology, and can be expressed in very abstract and general terms.

A Stable Population or Extinction?

Suppose a world in which organisms have many offspring. Since the offspring themselves will have many offspring, the population will grow and grow ("geometrically") until inevitably, sooner or later—surprisingly soon, in fact—it must grow too large for the available resources (of food, of space, of whatever the organisms need to survive long enough to reproduce). At that point, whenever it happens, not all organisms will have offspring. Many will die childless. It was Malthus who pointed out the mathematical inevitability of such a crunch in *any* population of long-term reproducers—people, animals, plants (or, for that matter, Martian clone-machines, not that such fanciful possibilities were discussed by Malthus). Those populations that reproduce at less than the replacement rate are headed for extinction unless they reverse the trend. Populations that maintain a stable population over long periods of time will do so by settling on a rate of overproduction of offspring that is balanced by the vicissitudes encountered. This is obvious, perhaps, for houseflies and other prodigious breeders, but Darwin drove the point home with a calculation of his own: "The elephant is reckoned to be the slowest breeder of all known animals, and I have taken some pains to estimate its probable minimum rate of natural increase: . . . at the end of the fifth century there would be alive fifteen million elephants, descended from the first pair." Since elephants have been around for millions of years, we can be sure that only a fraction of the elephants born in any period have progeny of their own.

So the normal state of affairs for any sort of reproducers is one in which more offspring are produced in any one generation than will in turn reproduce in the next. In other words, it is almost always crunch time. At such a crunch, which prospective parents will "win"? Will it be a fair lottery, in which every organism has an equal chance of being among the few that reproduce? In a political context, this is where invidious themes enter, about power, privilege, injustice, treachery, class warfare, and the like, but we can elevate the observation from its political birthplace and consider in the abstract, as Darwin did, what would—must—happen in nature. Darwin added two further logical points to the insight he had found in Malthus: the first was that at crunch time, if there was significant variation among the contestants, then any advantages enjoyed by any of the contestants would inevitably bias the sample that reproduced. However tiny the advantage in question, if it was actually an advantage (and thus not absolutely invisible to nature), it would tip the scales in favor of those who held it. The second was that *if* there was a "strong principle of inheritance"—if offspring tended to be more like their parents than like their parents' contemporaries—the biases created by advantages, however small, would become amplified over time, creating trends that could grow indefinitely. "More individuals are born than can possibly survive. A grain in the balance will determine which individual shall live and which shall die,—which variety or species shall increase in number, and which shall decrease, or finally become extinct."

Darwin's Dangerous Idea

What Darwin saw was that if one merely supposed these few general conditions to apply at crunch time—conditions for which he could supply ample evidence—the resulting process would *necessarily* lead in the direction of individuals in future generations who tended to be better equipped to deal with the problems of resource limitation that had been faced by the individuals of their parents' generation. This fundamental idea—Darwin's dangerous idea, the idea that generates so much insight, turmoil, confusion, anxiety—is thus actually quite simple. Darwin summarizes it in two long sentences at the end of chapter 4 of *Origin:*

> If during the long course of ages and under varying conditions of life, organic beings vary at all in the several parts of their organization, and I think this cannot be disputed; if there be, owing to the high geometric powers of increase of each species, at some age, season, or year, a severe struggle for life, and this certainly cannot be disputed; then, considering the infinite complexity of the relations of all organic beings to each other and to their conditions of existence, causing an infinite diversity in structure, constitution, and habits, to be advantageous to them, I think it would be a most extraordinary fact if no variation ever had occurred useful to each being's own welfare, in the same way as so many variations have occurred useful to man. But if variations useful to any organic being do occur, assuredly individuals thus characterized will have the best chance of being preserved in the struggle for life; and from the strong principle of inheritance they will tend to produce offspring similarly characterized. This principle of preservation, I have called, for the sake of brevity, Natural Selection.

This was Darwin's great idea, not the idea of evolution, but the idea of evolution *by natural selection,* an idea he himself could never formulate with sufficient rigor and detail to prove, though he presented a brilliant case for it.

Darwin Describes the "Struggle for Existence"

CHARLES DARWIN

Although secondhand summaries and descriptions of Charles Darwin's concept of natural selection are helpful in conveying how that process works, none can substitute for Darwin's own words, which remain a marvelous combination of logical organization and the concise, clear presentation of a complex subject. This excerpt from his Origin of Species *is from the conclusion, in which he gives a masterful synopsis of the main points that he has covered in more detail earlier. These include the concepts of natural favorable and unfavorable variations, competition among species, extinction, effects of disuse of organs and limbs, human domestication of plants and animals, and more.*

Under domestication we see much variability, caused, or at least excited, by changed conditions of life; but often in so obscure a manner, that we are tempted to consider the variations as spontaneous. Variability is governed by many complex laws,—by correlated growth, compensation, the increased use and disuse of parts, and the definite action of the surrounding conditions. There is much difficulty in ascertaining how largely our domestic productions have been modified; but we may safely infer that the amount has been large, and that modifications can be inherited for long periods. As long as the conditions of life remain the same,

Charles Darwin, *The Origin of Species*. New York: American Library, 1958.

we have reason to believe that a modification, which has already been inherited for many generations, may continue to be inherited for an almost infinite number of generations. On the other hand, we have evidence that variability when it has once come into play, does not cease under domestication for a very long period; nor do we know that it ever ceases, for new varieties are still occasionally produced by our oldest domesticated productions.

The Unconscious Process of Selection

Variability is not actually caused by man; he only unintentionally exposes organic beings to new conditions of life, and then nature acts on the organisation and causes it to vary. But man can and does select the variations given to him by nature, and thus accumulates them in any desired manner. He thus adapts animals and plants for his own benefit or pleasure. He may do this methodically, or he may do it unconsciously by preserving the individuals most useful or pleasing to him without any intention of altering the breed. It is certain that he can largely influence the character of a breed by selecting, in each successive generation, individual differences so slight as to be inappreciable except by an educated eye. This unconscious process of selection has been the great agency in the formation of the most distinct and useful domestic breeds. That many breeds produced by man have to a large extent the character of natural species, is shown by the inextricable doubts whether many of them are varieties or aboriginally distinct species.

There is no reason why the principles which have acted so efficiently under domestication should not have acted under nature. In the survival of favoured individuals and races, during the constantly-recurrent Struggle for Existence, we see a powerful and ever-acting form of Selection. The struggle for existence inevitably follows from the high geometrical ratio of increase which is common to all organic beings. This high rate of increase is proved by calculation,—by the rapid increase of many animals and plants during a succession of peculiar seasons, and when naturalised in new countries. More individuals are born than can possibly survive. A grain in the balance may determine which individuals shall live and which shall die,—which variety or species shall increase in number, and which shall decrease, or finally become extinct. As the individuals of the same species come in all respects into the closest competition with each other, the struggle will generally be most severe between them; it will be almost equally severe between the varieties of the same species, and next in severity between the species of the same genus. On the other hand the struggle will often be severe between beings remote in the scale of nature. The slightest advantage in certain individuals, at any age or during any season, over those with which they come into competition, or better adaptation in however slight a degree to the surrounding physical conditions, will, in the long run, turn the balance.

With animals having separated sexes, there will be in most cases a struggle between the males for the possession of the females. The most vigorous males, or those which have most successfully struggled with their conditions of life, will generally leave most progeny. But success will often depend on the males having special weapons, or means of defence, or charms; and a slight advantage will lead to victory.

As geology plainly proclaims that each land has undergone great physical changes, we might have expected to find that organic beings have varied under nature, in the same way as they have varied under domestication. And if there has been any variability under nature, it would be an unaccountable fact if natural selection had not come into play. It has often been asserted, but the assertion is incapable of proof, that the amount of variation under nature is a strictly limited quantity. Man, though acting on external characters alone and often capriciously, can produce within a short period a great result by adding up mere individual differences in his domestic productions; and every one admits that species present individual differences. But, besides such differences, all naturalists admit that natural varieties exist, which are considered sufficiently distinct to be worthy of record in systematic works. No one has drawn any clear distinction between individual differences and slight varieties; or between more plainly marked varieties and sub-species, and species. On separate continents, and on different parts of the same continent when divided by barriers of any kind, and on outlying islands, what a multitude of forms exist, which some experienced naturalists rank as varieties, others as geographical races or sub-species, and others as distinct, though closely allied species!

If then, animals and plants do vary, let it be ever so slightly or slowly, why should not variations or individual differences, which are in any way beneficial, be preserved and accumulated through natural selection, or the survival of the fittest? If man can by patience select variations useful to him, why, under changing and complex conditions of life, should not variations useful to nature's living products often arise, and be preserved or selected? What limit can be put to this power, acting during long ages and rigidly scrutinising the whole constitution, structure, and habits of each creature,—favouring the good and rejecting the bad? I can see no limit to this power, in slowly and beautifully adapting each form to the most complex relations of life. The theory of natural selection, even if we look no farther than this, seems to be in the highest degree probable. I have already recapitulated, as fairly as I could, the opposed difficulties and objections: now let us turn to the special facts and arguments in favour of the theory.

Accumulating Favourable Variations

On the view that species are only strongly marked and permanent varieties, and that each species first existed as a variety, we can see why it is that no line of demarcation can be drawn between species, commonly supposed to have been produced by special acts of creation, and varieties which are acknowledged to have been produced by secondary laws. On this same view we can understand how it is that in a region where many species of a genus have been produced, and where they now flourish, these same species should present many varieties; for where the manufactory of species has been active, we might expect, as a general rule, to find it still in action; and this is the case if varieties be incipient species. Moreover, the species of the larger genera, which afford the greater number of varieties or incipient species, retain to a certain degree the character of varieties; for they differ from each other by a less amount of difference than do the species of smaller genera. The closely allied species also of the larger genera apparently have restricted ranges, and in their affinities they are clustered in little groups round other species—in both respects resembling varieties. These are strange relations on the view that each species was independently created, but are intelligible if each existed first as a variety.

As each species tends by its geometrical rate of reproduction to increase inordinately in number; and as the modified descendants of each species will be enabled to increase by as much as they become more diversified in habits and structure, so as to be able to seize on many and widely different places in the economy of nature, there will be a constant tendency in natural selection to preserve the most divergent offspring of any one species. Hence, during a long-continued course of modification, the slight differences characteristic of varieties of the same species, tend to be augmented into the greater differences characteristic of the species of the same genus. New and improved varieties will inevitably supplant and exterminate the older, less improved, and intermediate varieties; and thus species are rendered to a large extent defined and distinct objects. Dominant species belonging to the larger groups within each class tend to give birth to new and dominant forms; so that each large group tends to become still larger, and at the same time more divergent in character. But as all groups cannot thus go on increasing in size, for the world would not hold them, the more dominant groups beat the less dominant. This tendency in the large groups to go on increasing in size and diverging in character, together with the inevitable contingency of much extinction, explains the arrangement of all the forms of life in groups subordinate to groups, all within a few great classes, which has prevailed throughout all time. This grand fact of the grouping of all organic beings under what is called the Natural System, is utterly inexplicable on the theory of creation.

As natural selection acts solely by accumulating slight, successive, favourable variations, it can produce no great or sudden modifications; it can act only by short and slow steps. Hence, the canon of "Natura non facit saltum" ["Nature makes no leaps"], which every fresh addition to our knowledge tends to confirm, is on this theory intelligible. We can see why throughout nature the same general end is gained by an almost infinite diversity of means, for every peculiarity when once acquired is long inherited, and structures already modified in many different ways have to be adapted for the same general purpose. We can, in short, see why nature is prodigal in variety, though niggard in innovation. But why this should be a law of nature if each species has been independently created no man can explain.

Nature Not Perfect

Many other facts are, as it seems to me, explicable on this theory. How strange it is that a bird, under the form of a woodpecker, should prey on insects on the ground; that upland geese which rarely or never swim, should possess webbed feet; that a thrush-like bird should dive and feed on sub-aquatic insects; and that a petrel should have the habits and structure fitting it for the life of an auk! and so in endless other cases. But on the view of each species constantly trying to increase in number, with natural selection always ready to adapt the slowly varying descendants of each to any unoccupied or ill-occupied place in nature, these facts cease to be strange, or might even have been anticipated.

We can to a certain extent understand how it is that there is so much beauty throughout nature; for this may be largely attributed to the agency of selection. That beauty, according to our sense of it, is not universal, must be admitted by every one who will look at some venomous snakes, at some fishes, and at certain hideous bats with a distorted resemblance to the human face. Sexual selection has given the most brilliant colours, elegant patterns, and other ornaments to the males, and sometimes to both sexes of many birds, butterflies, and other animals. With birds it has often rendered the voice of the male musical to the female, as well as to our ears. Flowers and fruit have been rendered conspicuous by brilliant colours in contrast with the green foliage, in order that the flowers may be readily seen, visited and fertilised by insects, and the seeds disseminated by birds. How it comes that certain colours, sounds, and forms should give pleasure to man and the lower animals,—that is, how the sense of beauty in its simplest form was first acquired,—we do not know any more than how certain odours and flavours were first rendered agreeable.

As natural selection acts by competition, it adapts and improves the inhabitants of each country only in relation to their co-inhabitants; so that we need feel no surprise at the species of any one country, although on the ordinary view supposed to have been created and specially adapted for that country, being beaten and supplanted by the naturalised productions from another land. Nor ought we to marvel if all the contrivances in nature be not, as far as we can judge, absolutely perfect, as in the case even of the human eye; or if some of them be abhorrent to our ideas of fitness. We need not marvel at the sting of the bee, when used against an enemy, causing the bee's own death; at drones being produced in such great numbers for one single act, and being then slaughtered by their sterile sisters; at the astonishing waste of pollen by our fir-trees; at the instinctive hatred of the queen-bee for her own fertile daughters; at the ichneumonidæ feeding within the living bodies of caterpillars; or at other such cases. The wonder indeed is, on the theory of natural selection, that more cases of the want of absolute perfection have not been detected. . . .

If we admit that the geological record is imperfect to an extreme degree, then the facts, which the record does give, strongly support the theory of descent with modification. New species have come on the stage slowly and at successive intervals; and the amount of change, after equal intervals of time, is widely different in different groups. The extinction of species and of whole groups of species which has played so conspicuous a part in the history of the organic world, almost inevitably follows from the principle of natural selection; for old forms are supplanted by new and improved forms. Neither single species nor groups of species reappear when the chain of ordinary generation is once broken. The gradual diffusion of dominant forms, with the slow modification of their descendants, causes the forms of life, after long intervals of time to appear as if they had changed simultaneously throughout the world. The fact of the fossil remains of each formation being in some degree intermediate in character between the fossils in the formations above and below, is simply explained by their intermediate position in the chain of descent. The grand fact that all extinct beings can be classed with all recent beings, naturally follows from the living and the extinct being the offspring of common parents. As species have generally diverged in character during their long course of descent and modification, we can understand why it is that the more ancient forms, or early progenitors of each group, so often occupy a position in some degree intermediate between existing groups. Recent forms are generally looked upon as being, on the whole, higher in the scale of organisation than ancient forms; and they must be higher, in so far as the later and more improved forms have conquered the older and less improved forms in the struggle for life; they have also generally had their organs more specialised for different functions. This fact is perfectly compatible with numerous beings still retaining simple but little improved structures, fitted for simple conditions of life; it is likewise compatible with some forms having retrograded in organisation, by having become at each stage of descent better fitted for new and degraded habits of life. Lastly, the wonderful law of the long endurance of allied forms on the same continent,—of marsupials in Australia, of edentata in America, and other such cases,—is intelligible, for within the same country the existing and the extinct will be closely allied by descent.

Looking to geographical distribution, if we admit that there has been during the long course of ages much migration from one part of the world to another, owing to former climatal and geographical changes and to the many occasional and unknown means of dispersal, then we can understand, on the theory of descent with modification, most of the great leading facts in Distribution. We can see why there should be so striking a parallelism in the distribution of organic beings throughout space, and in their geological succession throughout time; for in both cases the beings have been connected by the bond of ordinary generation, and the means of modification have been the same. We see the full meaning of the wonderful fact, which has struck every traveller, namely, that on the same continent, under the most diverse conditions, under heat and cold, on mountain and lowland, on deserts and marshes, most of the inhabitants within each great class are plainly related; for they are the descendants of the same progenitors and early colonists. On this same principle of former migration, combined in most cases with modification, we can understand, by the aid of the Glacial period, the identity of some few plants, and the close alliance of many others, on the most distant mountains, and in the northern and southern temperate zones; and likewise the close alliance of some of the inhabitants of the sea in the northern and southern temperate latitudes, though separated by the whole intertropical ocean. Although two countries may present physical conditions as closely similar as the same species ever require, we need feel no surprise at their inhabitants being widely different, if they have been for a long period completely sundered from each other; for as the relation of organism to organism is the most important of all relations, and as the two countries will have received colonists at various periods and in different proportions, from some other country or from each other, the course of modification in the two areas will inevitably have been different. . . .

How Disuse Affects Natural Selection

The similar framework of bones in the hand of a man, wing of a bat, fin of the porpoise, and leg of the horse—the same number of vertebræ forming the neck of the giraffe and of the elephant,—and innumerable other such facts, at once explain themselves on the theory of descent with slow and slight successive modifications. The similarity of pattern in the wing and in the leg of a bat, though used for such different purpose,—in the jaws and legs of a crab,—in the petals, stamens, and pistils of a flower, is likewise, to a large extent, intelligible on the view of the gradual modification of parts or organs, which were aboriginally alike in an early progenitor in each of these classes. On the principle of successive variations not always supervening at an early age, and being inherited at a corresponding not early period of life, we clearly see why the embryos of mammals, birds, reptiles, and fishes should be so closely similar, and so unlike the adult forms. We may cease marvelling at the embryo of an air-breathing mammal or bird having branchial slits and arteries running in loops, like those of a fish which has to breathe the air dissolved in water by the aid of well-developed branchiæ.

Disuse, aided sometimes by natural selection, will often have reduced organs when rendered useless under changed habits or conditions of life; and we can understand on this

view the meaning of rudimentary organs. But disuse and selection will generally act on each creature, when it has come
to maturity and has to play its full part in the struggle for existence, and will thus have little power on an organ during
early life; hence the organ will not be reduced or rendered
rudimentary at this early age. The calf, for instance, has inherited teeth, which never cut through the gums of the upper jaw, from an early progenitor having well-developed
teeth; and we may believe, that the teeth in the mature animal were formerly reduced by disuse, owing to the tongue
and palate, or lips, having become excellently fitted through
natural selection to browse without their aid; whereas in the
calf, the teeth have been left unaffected, and on the principle of inheritance at corresponding ages have been inherited from a remote period to the present day. On the view
of each organism with all its separate parts having been specially created, how utterly inexplicable is it that organs bearing the plain stamp of inutility, such as the teeth in the em-
bryonic calf or the shrivelled wings under the soldered
wing-covers of many beetles, should so frequently occur.
Nature may be said to have taken pains to reveal her
scheme of modification, by means of rudimentary organs,
of embryological and homologous structures, but we are
too blind to understand her meaning.

I have now recapitulated the facts and considerations
which have thoroughly convinced me that species have
been modified, during a long course of descent. This has
been effected chiefly through the natural selection of numerous successive, slight, favourable variations; aided in an
important manner by the inherited effects of the use and disuse of parts; and in an unimportant manner, that is in relation to adaptive structures, whether past or present, by the
direct action of external conditions, and by variations which
seem to us in our ignorance to arise spontaneously. . . .

It can hardly be supposed that a false theory would explain, in so satisfactory a manner as does the theory of natural selection.

Darwin's Unexpected Competitor

ALFRED RUSSEL WALLACE

Charles Darwin was not the only nineteenth-century thinker to develop a workable theory of evolutionary natural selection. While working on his book expounding his own theory, Darwin was surprised to learn that a younger naturalist, Alfred Russel Wallace, was independently working along extremely similar lines. Unaware of the scope of Darwin's work, Wallace sent him the following essay summarizing his own theory.

One of the strongest arguments which have been adduced to prove the original and permanent distinctness of species is, that *varieties* produced in a state of domesticity are more or less unstable, and often have a tendency, if left to themselves, to return to the normal form of the parent species; and this instability is considered to be a distinctive peculiarity of all varieties, even of these occurring among wild animals in a state of nature, and to constitute a provision for preserving unchanged the originally created distinct species.

In the absence or scarcity of facts and observations as to *varieties* occurring among wild animals, this argument has had great weight with naturalists, and has led to a very

John L. Brooks, *Just Before the Origin: Alfred Russel Wallace's Theory of Evolution.* New York: Columbia University Press, 1984.

general and somewhat prejudiced belief in the stability of species. Equally general, however, is the belief in what are called "permanent or true varieties,"—races of animals which continually propagate their like, but which differ so slightly (although constantly) from some other race, that the one is considered to be a *variety* of the other. Which is the *variety* and which the original *species*, there is generally no means of determining, except in those rare cases in which the one race has been known to produce an offspring unlike itself and resembling the other. This, however, would seem quite incompatible with the "permanent invariability of species," but the difficulty is overcome by assuming that such varieties have strict limits, and can never again vary further from the original type, although they may return to it, which, from the analogy of the domesticated animals, is considered to be highly probable, if not certainly proved.

It will be observed that this argument rests entirely on the assumption, that *varieties* occurring in a state of nature are in all respects analogous to or even identical with those of domestic animals, and are governed by the same laws as regards their permanence or further variation. But it is the object of the present paper to show that this assumption is altogether false, that there is a general principle in nature which will cause many *varieties* to survive the parent species, and to give rise to successive variations departing further and further from the original type, and which also produces, in domesticated animals, the tendency of varieties to return to the parent form.

Large and Small Animal Populations

The life of wild animals is a struggle for existence. The full exertion of all their faculties and all their energies is required to preserve their own existence and provide for that of their infant offspring. The possibility of procuring food during the least favourable seasons, and of escaping the attacks of their most dangerous enemies, are the primary conditions which determine the existence both of individuals and of entire species. These conditions will also determine the population of a species; and by a careful consideration of all the circumstances we may be enabled to comprehend, and in some degree to explain, what at first sight appears to be inexplicable—the excessive abundance of some species, while others closely allied to them are very rare.

The general proportion that must obtain between certain groups of animals is readily seen. Large animals cannot be so abundant as small ones; the carnivora [meat eaters] must be less numerous than the herbivora [plant eaters]: eagles and lions can never be so plentiful as pigeons and antelopes; the wild asses of the Tartarian deserts cannot equal in numbers the horses of the more luxuriant prairies and pampas of America. The greater or less fecundity [fertility] of an animal is often considered to be one of the chief causes of its abundance or scarcity; but a consideration of the facts will show us that it really has little or nothing to do with the matter. Even the least prolific of animals would increase rapidly if unchecked, whereas it is evident that the animal population of the globe must be stationary, or perhaps, through the influence of man, decreasing. Fluctuations there may be; but permanent increase, except in restricted localities, is almost impossible. For example, our own observation must convince us that birds do not go on increasing every year in a geometrical ratio, as they would do, were there not some powerful check to their natural increase. Very few birds produce less than two young ones each year, while many have six, eight, or ten; four will certainly be below the average; and if we suppose that each pair produce young only four times in their life, that will also be below the average, supposing them not to die either by violence or want of food. Yet at this rate how tremendous would be the increase in a few years from a single pair! A simple calculation will show that in fifteen years each pair of birds would have increased to nearly ten millions! whereas we have not reason to believe that the number of the birds of any country increases at all in fifteen or in one hundred and fifty years. With such powers of increase the population must have reached its limits, and have become stationary, in a very few years after the origin of each species. It is evident, therefore, that each year an immense number of birds perish—as many in fact as are born; and as on the lowest calculation the progeny are each year twice as numerous as their parents, it follows that, whatever be the average number of individuals existing in any given country, *twice that number must perish annually*—a striking result, but one which seems at least highly probable, and is perhaps under rather than over the truth.

The Strongest Outlive the Weakest

It would therefore appear that, as far as the continuance of the species and the keeping up the average numbers of individuals are concerned, large broods are superfluous. On the average all above *one* become food for hawks and kites, wild cats and weasels, or perish of cold and hunger as winter comes on. This is strikingly proved by the case of particular species; for we find that their abundance in individuals bears no relation whatever to their fertility in producing offspring. Perhaps the most remarkable instance of an immense bird population is that of the passenger pigeon of the United States which lays only one, or at most two eggs, and is said to rear generally but one young one. Why is this bird so extraordinarily abundant, while others producing two or three times as many young are much less plentiful? The explanation is not difficult. The food most congenial to this species, and on which it thrives best, is abundantly distributed over a very extensive region, offering such differences of soil and climate, that in one part or another of the area the supply never fails. The bird is capable of a very rapid and long-continued flight, so that it can pass without fatigue over the whole of the district it inhabits, and as soon as the supply of food begins to fail in one place is able to discover a fresh feeding ground. This example strikingly shows us that the procuring a constant supply of wholesome food is almost the sole condition requisite for ensuring the rapid increase of a given species, since neither the limited fecundity, nor the unrestrained attacks of birds of prey and of man are here sufficient to check it. In no other birds are these peculiar circumstances so strikingly combined. Either their food is more liable to failure, or they have not sufficient power of wing to search for it over an extensive area, or during some season of the year it becomes very scarce, and less wholesome substitutes have to be found; and thus, though more fertile in offspring, they can never increase beyond the supply of food in the least favourable seasons. . . .

This is probably the reason why woodpeckers are scarce with us, while in the tropics they are among the most abundant of solitary birds. Thus the house sparrow is more abundant than the redbreast, because its food is more constant and plentiful,—seeds of grasses being preserved during the winter, and our farm-yards and stubble-fields furnishing an almost inexhaustible supply. Why, as a general rule, are aquatic, and especially sea birds, very numerous in individuals? Not because they are more prolific than others, generally the contrary; but because their food never fails, the sea-shores and river-banks daily swarming with a fresh supply of small mollusca and crustacea. Exactly the same laws will apply to mammals. Wild cats are prolific and have few enemies; why then are they never as abundant as rabbits? The only intelligible answer is, that their supply of food is more precarious. It appears evident, therefore, that so long as a country remains physically unchanged, the numbers of its animal population cannot materially increase. If one species does so, some other requiring the same kind of food must diminish in proportion. The numbers that die annually must be immense; and as the individual existence of each animal depends upon itself, those that die must be the weakest—the very young, the aged, and the diseased,—while those that prolong their existence can only be the most perfect in health and vigour—those who are best able to obtain food regularly, and avoid their numerous enemies. It is, as we commenced by remarking, "a struggle for existence," in which the weakest and least perfectly organized must always succumb.

Variety in Species

Now it is clear that what takes place among the individuals of a species must also occur among the several allied species of a group,—viz. that those which are best adapted to obtain a regular supply of food, and to defend themselves against the attacks of their enemies and the vicissitudes of the seasons, must necessarily obtain and preserve a superiority in population; while those species which from some defect of power or organization are the least capable of counteracting the vicissitudes of food, supply, &c., must diminish in numbers, and, in extreme cases, become altogether extinct. Between these extremes the species will present various degrees of capacity for ensuring the means of preserving life; and it is thus we account for the abundance or rarity of species. Our ignorance will generally prevent us from accurately tracing the effects to their causes; but could we become perfectly acquainted with the organization and habits of the various species of animals, and could we measure the capacity of each for performing the different acts necessary to its safety and existence under all the varying circumstances by which it is surrounded, we might be able even to calculate the proportionate abundance of individuals which is the necessary result.

If now we have succeeded in establishing these two points—1st, *that the animal population of a country is generally stationary, being kept down by a periodical deficiency of food and other checks;* and, 2nd, *that the comparative abundance or scarcity of the individuals of the several species is entirely due to their organization and resulting*

habits, which, rendering it more difficult to procure a regular
supply of food and to provide for their personal safety in
some cases than in others, can only be balanced by a differ-
ence in the population which have to exist in a given area—
we shall be in a condition to proceed to the consideration
of *varieties*, to which the preceding remarks have a direct
and very important application.

Most or perhaps all the variations from the typical form
of a species must have some definite effect, however slight,
on the habits or capacities of the individuals. Even a
change of colour might, by rendering them more or less
distinguishable, affect their safety; a greater or less development of hair might modify their habits. More important
changes, such as an increase in the power or dimensions
of the limbs or any of the external organs, would more or
less affect their mode of procuring food or the range of
country which they inhabit. It is also evident that most
changes would affect, either favourably or adversely, the
powers of prolonging existence. An antelope with shorter
or weaker legs must necessarily suffer more from the attacks of the feline carnivora; the passenger pigeon with less
powerful wings would sooner or later be affected in its
powers of procuring a regular supply of food; and in both
cases the result must necessarily be a diminution of the
modified species. If, on the other hand, any species should
produce a variety having slightly increased powers of preserving existence, that variety must inevitably in time acquire a superiority in numbers. These results must follow
as surely as old age, intemperance, or scarcity of food produce an increased mortality. In both cases there may be
many individual exceptions; but on the average the rule
will invariably be found to hold good. All varieties will
therefore fall into two classes—those which under the
same conditions would never reach the population of the
parent species, and those which would in time obtain and
keep a numerical superiority. Now, let some alteration of
physical conditions occur in the district—a long period of

drought, a destruction of vegetation by locusts, the irruption of some new carnivorous animal seeking "pastures new"—any change in fact tending to render existence more difficult to the species in question, and tasking its utmost powers to avoid complete extermination; it is evident that, of all the individuals composing the species, those forming the least numerous and most feebly organized variety would suffer first, and, were the pressure severe, must soon become extinct. The same causes continuing in action, the parent species would next suffer, would gradually diminish in numbers, and with a recurrence of similar unfavourable conditions might also become extinct. The superior variety would then alone remain, and on a return to favourable circumstances would rapidly increase in numbers and occupy the place of the extinct species and variety.

Rise of New Varieties

The *variety* would now have replaced the *species*, of which it would be a more perfectly developed and more highly organized form. It would be in all respects better adapted to secure its safety, and to prolong its individual existence and that of the race. Such a variety *could not* return to the original form; for that form is an inferior one, and could never compete with it for existence. Granted, therefore, a "tendency" to reproduce the original type of the species, still the variety must ever remain preponderant in numbers, and under adverse physical conditions *again alone survive.* But this new, improved, and populous race might itself, in course of time, give rise to new varieties, exhibiting several diverging modifications of form, any of which, tending to increase the facilities for preserving existence, must, by the same general law, in their turn become predominant. Here, then, we have *progression and continued divergence* deduced from the general laws which regulate the existence of animals in a state of nature, and from the undisputed fact that varieties do frequently occur. It is not, however, contended that this result would be invariable; a change of physical conditions in the district might at times materially modify it, rendering the race which had been the most capable of supporting existence under the former conditions now the least so, and even causing the extinction of the new and, for a time, superior race, while the old parent species and its first inferior varieties continued to flourish. Variations in unimportant parts might also occur, having no perceptible effect on the life-preserving powers; and the varieties so furnished might run a course parallel with the parent species, either giving rise to further variations or returning to the former type. All we argue for is, that certain varieties have a tendency to maintain their existence longer than the original species, and this tendency must make itself felt; for though the doctrine of chances or averages can never be trusted to on a limited scale, yet if applied to high numbers, the results come nearer to what theory demands, and, as we approach to an infinity of examples, become strictly accurate. Now the scale on which nature works is so vast—the numbers of individuals and periods of time with which she deals approach so near to infinity, that any cause, however slight, and however liable to be veiled and counteracted by accidental circumstances, must in the end produce its full legitimate results.

The Case of Domestic Animals

Let us now turn to domesticated animals, and inquire how varieties produced among them are affected by the principles here enunciated. The essential difference in the condition of wild and domestic animals is this,—that among the former, their well-being and very existence depend upon the full exercise and healthy condition of all their senses and physical powers, whereas among the latter, these are only partially exercised, and in some cases are absolutely unused. A wild animal has to search, and often to labour, for every mouthful of food—to exercise sight, hearing, and smell in seeking it, and in avoiding dangers, in procuring shelter from the inclemency of the seasons, and in providing for the subsistence and safety of its offspring. There is no muscle of its body that is not called into daily and hourly activity; there is no sense or faculty that is not strengthened by continual exercise. The domestic animal, on the other hand, has food provided for it, is sheltered, and often confined, to guard it against the vicissitudes of the seasons, is carefully secured from the attacks of its natural enemies, and seldom even rears its young without human assistance. Half of its senses and faculties are quite useless; and the other half are but occasionally called into feeble exercise, while even its muscular system is only irregularly called into action.

Now when a variety of such an animal occurs, having increased power or capacity in any organ or sense, such increase is totally useless, is never called into action, and may even exist without the animal ever becoming aware of it. In the wild animal, on the contrary, all its faculties and powers being brought into full action for the necessities of existence, any increase becomes immediately available, is strengthened by exercise, and must even slightly modify the food, the habits, and the whole economy of the race. It creates as it were a new animal, one of superior powers, and which will necessarily increase in numbers and outlive those inferior to it.

Again, in the domesticated animal all varieties have an equal chance of continuance; and those which would decidedly render a wild animal unable to compete with its fellows and continue its existence are no disadvantage whatever in a state of domesticity. Our quickly fattening pigs, short-legged sheep, pouter pigeons, and poodle dogs could never have come into existence in a state of nature, because the very first steps toward such inferior forms would have led to the rapid extinction of the race; still less could they now exist in competition with their wild allies. The great speed but slight endurance of the race horse, the unwieldy strength of the ploughman's team, would both be useless in a state of nature. If turned wild on the pampas, such animals would probably soon become extinct, or under favourable circumstances might each lose those extreme qualities which would never be called into action, and in a few generations would revert to a common type, which must be that in which the various powers and faculties are so proportioned to each other as to be best adapted to procure food and secure safety,—that in which by the full exercise of every part of his organization the animal can alone continue to live. Domestic varieties, when turned wild, *must* return to something near the type of the original wild stock, *or become altogether extinct.* . . .

Nature's Balance

The hypothesis of [Jean-Baptiste de Monet de] Lamarck—that progressive changes in species have been produced by the attempts of animals to increase the development of their own organs, and thus modify their structure and habits—has been repeatedly and easily refuted by all writers on the subject of varieties and species, and it seems to have been considered that when this was done the whole question has been finally settled; but the view here developed renders such an hypothesis quite unnecessary, by showing that similar results must be produced by the action of principles constantly at work in nature. The powerful retractile talons of the falcon- and the cat-tribes have not been produced or increased by the volition of those animals; but among the different varieties which occurred in the earlier and less highly organized forms of these groups, *those always survived longest which had the greatest facilities for seizing their prey.* Neither did the giraffe acquire its long neck by desiring to reach the foliage of the more lofty shrubs, and constantly stretching its neck for the purpose, but because any varieties which occurred among its antitypes [ancestral forms] with a longer neck than usual *at once secured a fresh range of pasture over the same ground as their shorter-necked companions, and on the first scarcity of food were thereby enabled to outlive them.*

Even the peculiar colours of many animals, especially insects, so closely resembling the soil or the leaves of the trunks on which they habitually reside, are explained on the same principle; for though in the course of ages varieties of many tints have occurred, *yet those races having colours best adapted to concealment from their enemies would inevitably survive the longest.* We have also here an acting cause to account for that balance so often observed in nature,—a deficiency in one set of organs always being compensated by an increased development of some others—powerful wings accompanying weak feet, or great velocity making up for the absence of defensive weapons; for it has been shown that all varieties in which an unbalanced deficiency occurred could not long continue their existence. The action of this principle is exactly like that of the centrifugal governor of the steam engine, which checks and corrects any irregularities almost before they become evident; and in like manner no unbalanced deficiency in the animal kingdom can ever reach any conspicuous magnitude, because it would make itself felt at the very first step, by rendering existence difficult and extinction almost sure soon to follow. . . .

We believe we have now shown that there is a tendency in nature to the continued progression of certain classes of *varieties* further and further from the original type—a progression to which there appears no reason to assign any definite limits—and that the same principle which produces this result in a state of nature will also explain why domestic varieties have a tendency to revert to the original type. This progression, by minute steps, in various directions, but always checked and balanced by the necessary conditions, subject to which alone existence can be preserved, may, it is believed, be followed out so as to agree with all the phenomena presented by organized beings, their extinction and succession in past ages, and all the extraordinary modifications of form, instinct, and habits which they exhibit.

A Contemporary Scientist Rejects Darwin's Theory

RICHARD OWEN

Charles Darwin's theory of evolution was criticized by some scientists when it first appeared in 1859. Among the most outspoken critics was Richard Owen, a distinguished English zoologist, who penned the following negative review of Darwin's book in the Edinburgh Review *in April 1860. The review begins with an overview of the main tenets of Darwin's theory. Owen then launches his attack. He claims that Darwin has not presented much in the way of evidence to back up his evolutionary ideas and instead expects the reader to trust in his intelligence and native ability to discover the truth. Also, says Owen, the coexistence of very simple and very complex lifeforms on Earth today seems to refute the theory of natural selection—for if simple creatures were always evolving into more complex forms, why are the most numerous creatures in nature the simplest in form?*

The interdependencies of living beings of different kinds and grades, and the injurious results of their interruption, have long attracted the attention of observant and philosophic naturalists. An undue importance indeed was at one time attached to this principle; it was deemed to be so absolute as that no one species could be permit-

Richard Owen, "Darwin on the Origin of Species," *Edinburgh Review*, April 7, 1860.

ted to perish without endangering the whole fabric of organization. . . .

Manifold subsequent experience has led to a truer appreciation and a more moderate estimate of the importance of the dependence of one living being upon another. Mr. Darwin contributes some striking and ingenious instances of the way in which the principle partially affects the chain, or rather network of life, even to the total obliteration of certain meshes. And truly extinction has made wide rents in the reticulation as now represented by the co-affinities of living species! . . .

[The] chief part [of Darwin's book], however, is devoted to speculations on the origin of species; and its main object is the advocacy of a view, which we find most clearly expressed in the following passage. Mr. Darwin refers to the multitude of the individual of every species, which, from one cause or another, perish either before, or soon after attaining maturity.

'Owing to this struggle for life, any variation, however slight and from whatever cause proceeding, if it be in any degree profitable to an individual of any species, in its infinitely complex relations to other organic beings and to external nature, will tend to the preservation of that individual, and will generally be inherited by its offspring. The offspring, also, will thus have a better chance of surviving, for, of the many individuals of any species which are periodically born, but a small number can survive. I have called this principle, by which each slight variation, if useful, is preserved, by the term of Natural Selection, in order to mark its relation to man's power of selection. We have seen that man by selection can certainly produce great results, and can adapt organic beings to his own uses, through the accumulation of slight but useful variations, given to him by the hand of Nature. But Natural Selection, as we shall hereafter see, is a power incessantly ready for action, and is as immeasurably superior to man's feeble efforts, as the works of Nature are to those of Art.'

Inadequate and Disappointing

The scientific world has looked forward with great interest to the facts which Mr. Darwin might finally deem adequate to the support of his theory on this supreme question in biology, and to the course of inductive original research which might issue in throwing light on 'that mystery of mysteries.' But having now cited the chief, if not the whole, of the original observations adduced by its author in the volume now before us, our disappointment may be conceived. Failing the adequacy of such observations, not merely to carry conviction, but to give a color to the hypothesis, we were then left to confide in the superior grasp of mind, strength of intellect, clearness and precision of thought and expression, which raise one man so far above his contemporaries, as to enable him to discern in the common stock of facts, of coincidences, correlations and analogies in Natural History, deeper and truer conclusions than his fellow-laborers had been able to reach.

These expectations, we must confess, received a check on perusing the first sentence in the book.

> 'When on board H.M.S. "Beagle," as naturalist, I was much struck with certain facts in the distribution of the inhabitants of South America, and in the geological relations of the present to the past inhabitants of that continent. These facts seemed to me to throw some light on the origin of species—that mystery of mysteries, as it has been called by some of our greatest philosophers.'

What is there, we asked ourselves, as we closed the volume to ponder on this paragraph,—what can there possibly be in the inhabitants, we suppose he means aboriginal inhabitants, of South America, or in their distribution on that continent, to have suggested to any mind that man might be a transmuted ape, or to throw any light on the origin of the human or other species? Mr. Darwin must be aware of what is commonly understood by an 'uninhabited island'; he may, however, mean by the inhabitants of South

America, not the human kind only, whether aboriginal or otherwise, but all the lower animals. Yet again, why are the freshwater polyps or sponges to be called 'inhabitants' more than the plants? Perhaps what was meant might be, that the distribution and geological relations of the organized beings generally in South America, had suggested transmutational [evolutionary] views. They have commonly suggested ideas as to the independent origin of such localized kinds of plants and animals. But what the 'certain facts' were, and what may be the nature of the light which they threw upon the mysterious beginning of species, is not mentioned or further alluded to in the present work.

The origin of species is the question of questions in Zoology; the supreme problem which the most striking of our original laborers, the clearest zoological thinkers, and the most successful generalists, have never lost sight of, whilst they have approached it with due reverence. We have a right to expect that the mind proposing to treat of, and assuming to have solved, the problem, should show its equality to the task. The signs of such intellectual power we look for in clearness of expression, and in the absence of all ambiguous or unmeaning terms. Now, the present work is occupied by arguments, beliefs, and speculations on the origin of species, in which, as it seems to us, the fundamental mistake is committed, of confounding the questions, of species being the result of a secondary cause or law, and of the nature of that creative law. . . .

Mr. Darwin rarely refers to the writings of his predecessors, from whom, rather than from the phenomena of the distribution of the inhabitants of South America, he might be supposed to have derived his ideas as to the origin of species. When he does allude to them, their expositions on the subject are inadequately represented. . . .

The Graveyards of Strata

Do the facts of actual organic nature square with the Darwinian hypothesis? Are all the recognized organic forms of the present date, so differentiated, so complex, so superior to conceivable primordial simplicity of form and structure, as to testify to the effects of Natural Selection continuously operating through untold time? Unquestionably not. The most numerous living beings now on the globe are precisely those which offer such a simplicity of form and structure, as best agrees, and we take leave to affirm can only agree, with that ideal prototype from which, by any hypothesis of natural law, the series of vegetable and animal life might have diverged.

If by the patient and honest study and comparison of plants and animals, under their manifold diversities of matured form, and under every step of development by which such form is attained, any idea may be gained of a hypothetical primitive organism,—if its nature is not to be left wholly to the unregulated fancies of dreamy speculation—we should say that the form and condition of life which are common, at one period of existence, to every known kind and grade of organism, would be the only conceivable form and condition of the one primordial being from which 'Natural Selection' infers that all the organisms which have ever lived on this earth have descended. . . .

Lasting and fruitful conclusions have, indeed, hitherto [before this] been based only on the possession of knowledge; now we are called upon to accept an hypothesis on the plea of want of knowledge. The geological record, it is averred, is so imperfect! But what human record is not? Especially must the record of past organisms be much less perfect than of present ones. We freely admit it. But when Mr. Darwin, in reference to the absence of the intermediate fossil forms required by his hypothesis—and only the zootomical zoologist can . . . appreciate their immense numbers—the countless hosts of transitional links which, on 'natural selection,' must certainly have existed at one period or another of the world's history—when Mr. Darwin exclaims what may be, or what may not be, the forms yet forthcoming out of the graveyards of strata, we would reply, that our only ground for prophesying of what may come, is by the analogy of what has come to light. . . .

The essential element in the complex idea of species, as it has been variously framed and defined by naturalists, viz. [concerning] the blood-relationship between all the individuals of such species, is annihilated on the hypothesis of 'natural selection.' According to this view a genus, a family, an order, a class, a sub-kingdom,—the individuals severally representing these grades of difference or relationship,—now differ from individuals of the same species only in degree: the species, like every other group, is a mere creature of the brain; it is no longer from nature. With the present evidence from form, structure, and procreative phenomena, of the truth of the opposite proposition, that 'classification is the task of science, but species the work of nature,' we believe that this aphorism will endure; we are certain that it has not yet been refuted.

Darwin's Supporters and Critics Clash in Public

FRANCIS DARWIN (AND VARIOUS EYEWITNESSES)

In 1892, Charles Darwin's son, Francis, published a volume of his father's letters and other documents relating to his life and work. Included was a collection of eyewitness descriptions of the now famous public meeting of the British Association for the Advancement of Science that took place at Oxford University on June 28, 1860. That evening, several public figures who vehemently opposed Darwin's theory of evolution, among them the noted scientist Richard Owen, gathered with the intention of attacking Darwin's theory and ruining his reputation as a scientist. When the meeting began, the leader of the anti-Darwinians, the popular bishop Samuel Wilberforce, spoke eloquently but failed to achieve his goal of destroying Darwin's credibility. Instead, Wilberforce showed that he did not understand the theory of evolution and made a fool of himself. Meanwhile, Darwin's major supporters, leading scientists Joseph D. Hooker and Thomas H. Huxley, won the day by convincing the vast majority of those gathered that Darwin's theory at least deserved a fair hearing.

T he meeting of the British Association at Oxford in 1860 is famous for two pitched battles over the *Origin of Species*. Both of them originated in unimportant papers. On Thursday, June 28th, Dr. Daubeny of Oxford made a com-

---

Francis Darwin, ed., *Charles Darwin: His Life Told in an Autobiographical Chapter and in a Selected Series of His Published Letters.* New York: Appleton, 1892.

munication to Section D: "On the final causes of the sexuality of plants, with particular reference to Mr. Darwin's work on the *Origin of Species.*" Mr. [Thomas H.] Huxley was called on by the President, but tried (according to the *Athenœum* report) to avoid a discussion, on the ground "that a general audience, in which sentiment would unduly interfere with intellect, was not the public before which such a discussion should be carried on." However, the subject was not allowed to drop. Sir R. Owen (I quote from the *Athenœum*, July 7th, 1860), who "wished to approach this subject in the spirit of the philosopher," expressed his "conviction that there were facts by which the public could come to some conclusion with regard to the probabilities of the truth of Mr. Darwin's theory." He went on to say that the brain of the gorilla "presented more differences, as compared with the brain of man, than it did when compared with the brains of the very lowest and most problematical of the Quadrumana." Mr. Huxley replied, and gave these assertions a "direct and unqualified contradiction," pledging himself to "justify that unusual procedure elsewhere," a pledge which he amply fulfilled. On Friday there was peace, but on Saturday 30th, the battle arose with redoubled fury, at a conjoint meeting of three Sections, over a paper by Dr. Draper of New York, on the "Intellectual development of Europe considered with reference to the views of Mr. Darwin."

An Empty Speech

The following account is from an eye-witness of the scene.

"The excitement was tremendous. The Lecture-room, in which it had been arranged that the discussion should be held, proved far too small for the audience, and the meeting adjourned to the Library of the Museum, which was crammed to suffocation long before the champions entered the lists. The numbers were estimated at from 700 to 1000. Had it been term-time, or had the general public been admitted, it would have been impossible to have accommodated the rush to hear the oratory of the bold Bishop [Samuel Wilberforce]. Prof. Henslow, the President of Section D, occupied the chair, and wisely announced . . . that none who had not valid arguments to bring forward on one side or the other, would be allowed to address the meeting: a caution that proved necessary, for no fewer than four combatants had their utterances burked by him, because of their indulgence in vague declamation.

"The Bishop was up to time, and spoke for full half-an-hour with inimitable spirit, emptiness and unfairness. It was evident from his handling of the subject that he had been 'crammed' up to the throat, and that he knew nothing at first hand; in fact, he used no argument not to be found in his *Quarterly Review* article. He ridiculed Darwin badly, and Huxley savagely, but all in such dulcet tones, so persuasive a manner, and in such well-turned periods, that I who had been inclined to blame the President for allowing a discussion that could serve no scientific purpose, now forgave him from the bottom of my heart."

Descent from a Monkey?

What follows is from notes most kindly supplied by the Hon. and Rev. W.H. Fremantle, who was an eye-witness of the scene.

"The Bishop of Oxford attacked Darwin, at first playfully but at last in grim earnest. It was known that the Bishop had written an article against Darwin in the last *Quarterly Review:* it was also rumoured that Prof. [Richard] Owen had been staying at Cuddesden and had primed the Bishop, who was to act as mouthpiece to the great Palæontologist, who did not himself dare to enter the lists [contest]. The Bishop, however, did not show himself master of the facts, and made one serious blunder. A fact which had been much dwelt on as confirmatory of Darwin's idea of variation, was that a sheep had been born shortly before in a flock in the North of England, having an addition of one to the vertebræ of the spine. The Bishop was declaring with rhetorical exaggeration that there was hardly any actual evidence on Darwin's side. 'What have they to bring forward?' he exclaimed. 'Some rumoured statement about a long-legged sheep.' But he passed on to banter: 'I should like to ask Professor Huxley, who is sitting by me, and is about to tear me to pieces when I have sat down, as to his belief in being descended from an ape. Is it on his grandfather's or his grandmother's side that the ape ancestry comes in?' And then taking a graver tone, he asserted in a solemn peroration that Darwin's views were contrary to the revelations of God in the Scriptures. Professor Huxley was unwilling to respond: but he was called for and spoke with his usual incisiveness and with some scorn. 'I am here only in the interest of science,' he said, 'and I have not heard anything which can prejudice the case of my august client.' Then after showing how little competent the Bishop was to enter upon the discussion, he touched on the question of Creation. 'You say that development [change, evolution] drives out the Creator. But you assert that God made you; and yet you know that you yourself were originally a little piece of matter no bigger than the end of this gold pencil-case.' Lastly as to the descent from a monkey, he said: 'I should feel it no shame to have risen from such an origin. But I should feel it a shame to have sprung from one who prostituted the gifts of culture and of eloquence to the service of prejudice and of falsehood.'

Not Ashamed of My Ancestors

"Many others spoke. Mr. Gresley, an old Oxford don, pointed out that in human nature at least orderly development was not the necessary rule; Homer was the greatest of poets, but he lived 3000 years ago, and has not produced his like.

"Admiral [Robert] Fitz-Roy [skipper of the *Beagle*, the ship that had carried Darwin on his voyage of discovery years before] was present, and said that he had often expostulated with his old comrade of the *Beagle* for entertaining views which were contradictory to the First Chapter of Genesis.

"Sir John Lubbock declared that many of the arguments by which the permanence of species was supported came to nothing, and instanced some wheat which was said to have come off an Egyptian mummy and was sent to him to prove that wheat had not changed since the time of the Pharaohs; but which proved to be made of French chocolate. Sir Joseph (then Dr.) Hooker spoke shortly, saying that he had found the hypothesis of Natural Selection so helpful in explaining the phenomena of his own subject of Botany, that he had been constrained to accept it. After a few words from Darwin's old friend Professor Henslow who occupied the chair, the meeting broke up, leaving the impression that those most capable of estimating the arguments of Darwin in detail saw their way to accept his conclusions."

Many versions of Mr. Huxley's speech were current: the following report of his conclusion is from a letter addressed by the late John Richard Green, then an undergraduate, to a fellow-student, now Professor Boyd Dawkins:—"I asserted, and I repeat, that a man has no reason to be ashamed of having all ape for his grandfather. If there were an ancestor whom I should feel shame in recalling, it would be a *man*, a man of restless and versatile intellect, who, not content with an equivocal success in his own sphere of activity, plunges into scientific questions with which he has no real acquaintance, only to obscure them by an aimless rhetoric, and distract the attention of his hearers from the real point at issue by eloquent digressions, and skilled appeals to religious prejudice."

The Theory of Social Darwinism

HERBERT SPENCER

Among the strongest nineteenth-century supporters of Darwin's theory of evolution, particularly the idea of "survival of the fittest," were thinkers and writers who ascribed to the concept of what became known as "social Darwinism." In essence, this view contends that evolutionary ideas such as the fittest surviving and entities adapting to changing surroundings can be applied not only to animals and plants over geologic time, but also to the social, political, and psychological development of human beings within their ever-changing society. Most scholars credit the seminal statement of social Darwinism to British sociologist and philosopher Herbert Spencer (1820–1903). He was already thinking about the concepts of natural selection and adaptation within a social context before Darwin published The Origin of Species *and felt that Darwin's work strongly corroborated his own. The following summary of Spencer's basic evolutionary ideas appeared in the* Westminster Review *in 1857. In this excerpt, Spencer argues that similar to organisms in nature, society evolves from homogenous to heterogenous forms. This process has led to the formation of social classes and the division of labor, he concludes.*

T he current conception of Progress is somewhat shifting and indefinite. Sometimes it comprehends little more than simple growth—as of a nation in the number of its

---

Herbert Spencer, "Progress: Its Law and Causes," *Westminster Review*, April 1857.

members and the extent of territory over which it has spread. Sometimes it has reference to quantity of material products—as when the advance of agriculture and manufactures is the topic. Sometimes the superior quality of these products is contemplated; and sometimes the new or improved appliances by which they are produced. When again we speak of moral or intellectual progress, we refer to the state of the individual or people exhibiting it; whilst, when the progress of Knowledge, of Science, of Art, is commented upon, we have in view certain abstract results of human thought and action. Not only, however, is the current conception of Progress more or less vague, but it is in great measure erroneous. It takes in not so much the reality of Progress as its accompaniments—not so much the substance as the shadow. That progress in intelligence which takes place during the evolution of the child into the man, or the savage into the philosopher, is commonly regarded as consisting in the greater number of facts known and laws understood: whereas the actual progress consist in the produce of a greater quantity and variety of articles for the satisfaction of men's wants; in the increasing security of person and property; in the widening freedom of action enjoyed whereas, rightly understood, social progress consists in those changes of structure in the social organism which have entailed these consequences. . . . The phenomena are contemplated solely as bearing on human happiness. Only those changes are held to constitute progress which directly or indirectly tend to heighten human happiness. And they are thought to constitute progress simply *because* they tend to heighten human happiness. But rightly to understand Progress, we must inquire what is the nature of these changes, considered apart from our interests. Ceasing, for example, to regard the successive geological modifications that have taken place in the Earth, as modifications that have gradually fitted it for the habitation of Man, and as therefore a geological progress, we must seek to determine the character common to these modifications—the law to which they all conform. And similarly in every other case. Leaving out of sight concomitants and beneficial consequences, let us ask what Progress is in itself.

The Law of Organic Progress

In respect to that progress which individual organisms display in the course of their evolution, this question has been answered by the Germans. The investigations of [the natural scientists] Wolff, Goethe, and Van Baer have established the truth that the series of changes gone through during the development of a seed into a tree, or an ovum into an animal, constitute an advance from homogeneity of structure to heterogeneity of structure. In its primary stage, every germ consists of a substance that is uniform throughout, both in texture and chemical composition. The first step in its development is the appearance of a difference between two parts of this substance; or, as the phenomenon is described in physiological language—a differentiation. Each of these differentiated divisions presently begins itself to exhibit some contrast of parts; and by these secondary differentiations become as definite as the original one. This progress is continuously repeated—is simultaneously going on in all parts of the growing embryo; and by endless multiplication of these differentiations there is ultimately produced that complex combination of tissues and organs constituting the adult animal or plant. This is the course of evolution followed by all organisms whatever. It is settled beyond dispute that organic progress consists in a change from the homogeneous to the heterogeneous.

Now, we propose in the first place to show, that this law of organic progress is the law of all progress. Whether it be in the development of the Earth, in the development of Life upon its surface, the development of Society, of Government, of Manufactures, of Commerce, of Language, Literature, Science, Art, this same evolution of the simple into the complex, through a process of continuous differentiation, holds throughout. From the earliest traceable cosmical changes down to the latest results of civilization, we shall find that the transformation of the homogeneous into the heterogeneous, is that in which Progress essentially consists. . . .

Whether an advance from the homogeneous to the heterogeneous is or is not displayed in the biological history of the globe, it is clearly enough displayed in the progress of the latest and most heterogeneous creature—Man. It is alike true that, during the period in which the Earth has been peopled, the human organism has become more heterogeneous among the civilized divisions of the species and that the species, as a whole, has been growing more heterogeneous in virtue of the multiplication of races and the differentiation of these races from each other. . . .

Distinct Classes and Orders of Workers

In the course of ages, there arises, as among ourselves, a highly complex political organization of monarch, ministers, lords and commons, with their subordinate administrative departments, courts of justice, revenue offices, &c., supplemented in the provinces by municipal governments, county governments, parish or union governments—all of them more or less elaborated. By its side there grows up a highly complex religious organization, with its various grades of officials from archbishops down to sextons, its colleges, convocations, ecclesiastical courts, &c.; to all which must be added the ever-multiplying independent sects, each with its general and local authorities. And at the same time there is developed a highly complex aggregation of customs, manners, and temporary fashions, enforced by society at large, and serving to control those minor transactions between man and mar which are not regulated by civil and religious law. Moreover it is to be observed that this ever-increasing heterogeneity in the governmental appliances of each nation, has been accompanied by an increasing heterogeneity in the governmental appliances of different nations all of which are more or less unlike in their political systems and legislation in their creeds and religious institutions, in their customs and ceremonial usages.

Simultaneously there has been going on a second differentiation of a still more familiar kind; that, namely, by which the mass of the community has become segregated into distinct classes and orders of workers. While the governing part has been undergoing the complex development above described, the governed part has been undergoing an equally complex development, which has resulted in that minute division of labour characterizing advanced nations. It is needless to trace out this progress from its first stages, up through the caste divisions of the East and the incorporated guilds of Europe, to the elaborate producing and distributing organization existing among ourselves. Political economists have made familiar to all, the evolution which, beginning with a tribe whose members severally perform the same actions each for himself, ends with a civilized community whose members severally perform different actions for each other; and they have further explained the evolution through which the solitary producer of any one commodity, is transformed into a combination of producers who united under a master, take separate parts in the manufacture of such commodity. But there are yet other and higher phases of this advance from the homogeneous to the heterogeneous in the industrial structure of the social organism. Long after considerable progress has been made in the division of labour among different classes of workers, there is still little or no division of labour among the widely separated parts of the community: the nation continues comparatively homogeneous in the respect that in each district the same occupations are pursued. But when roads and other means of transit become numerous and good, the different districts begin to assume different functions, and to become mutually dependent. The calico manufacture locates it self in this county, the woollencloth manufacture in that; silks are produced here, lace there; stockings in one place, shoes in another; pottery, hardware, cutlery, come to have their special towns; and ultimately every locality becomes more or less distinguished from the rest by the leading occupation carried on in it. Nay, more, this subdivision of functions shows itself not only among the different parts of the same nation, but among different nations. That exchange of commodities which freetrade promises so greatly to increase, will ultimately have the effect of specializing, in a greater or less degree, the industry of each people. So that beginning with a barbarous tribe, almost if not quite homogeneous in the functions of its members, the progress has been, and still is, towards an economic aggregation of the whole human race, growing ever more heterogeneous in respect of the separate functions assumed by separate nations, the separate functions assumed by the local sections of each nation, the separate functions assumed by the many kinds of makers and traders in each town, and the separate functions assumed by the workers united in producing each commodity.

Not only is the law thus clearly exemplified in the evolution of the social organism, but it is exemplified with equal clearness in the evolution of all products of human thought and action; whether concrete or abstract, real or ideal. . . .

Evolution Happening at All Levels

We might trace out the evolution of Science; beginning with the era in which it was not yet differentiated from Art, and was, in union with Art, the handmaid of Religion; passing through the era in which the sciences were so few and rudimentary, as to be simultaneously cultivated by the same philosophers; and ending with the era in which the genera and species are so numerous that few can enumerate them, and no one can adequately grasp even one genus. Or we might do the like with Architecture, with the Drama, with Dress. But doubtless the reader is already weary of illustrations; and our promise has been amply fulfilled. We believe we have shown beyond question, that that which the German physiologists have found to be the law of organic development, is the law of all development. The advance from the simple to the complex, through a process of successive differentiations, is seen alike in the earliest changes of the Universe to which we can reason our way back, and in the earliest changes which we can inductively establish; it is seen in the geologic and climatic evolution of the Earth, and of every single organism on its surface; it is seen in the evolution of Humanity, whether contemplated in the civilized individual, or in the aggregation of races; it is seen in the evolution of Society in respect both of its political and economical organization; and it is seen in the evolution of all those endless concrete and abstract products of human activity which constitute the environment of our daily life. From the remotest past which Science can fathom, down to the novelties of yesterday, that in which Progress essentially consists, is the transformation of the homogeneous into the heterogeneous.

THE
HISTORY
OF
ISSUES

CHAPTER 3

Modern Science and Creationism at Odds

Chapter Preface

The modern controversy over the issue of evolution is based in large degree on people's trust in the ability of modern science to unravel the mysteries of the universe. Some people are content to allow scientists to do their work, invent and disprove theories, and slowly guide humanity to new levels of understanding. Others are not so trusting of science's potential and motives. They worry that some scientists have personal or political agendas that might lead them to back and promote some theories that are only partially or even poorly supported by evidence.

This difference in attitude about science and its potential for good or ill is a major factor in the ongoing controversy and debate between those who accept the doctrine of the evolution of living things and those who hold that God created life miraculously. Each side is positive that its basic tenets are correct and will eventually prevail. And each side is typically dismissive, sometimes even contemptuous, of the other.

One thing is certain. The two sides are not likely to come to a mutual understanding any time soon. A great deal of scientific research about evolution continues; and books and articles reporting the newest discoveries and hypotheses appear on a regular basis. Meanwhile, creationist literature criticizing and/or denouncing evolution has exploded in quantity in recent years. The Internet has become particularly fertile ground, a public forum in which many thousands of articles on the evolution/creationism debate can be found.

By far the most heated disagreement in the general debate among modern evolutionists and creationists involves the manner in which children should learn about the creation of living things. (Public schools are at the core of the issue since parents have more of a say in deciding the content of curricula in private schools.) Most scientists and many educated laypersons feel that only scientific doctrines like evolution should be taught in public schools. So-called creation science, they say, is not science but rather religion pretending to be science. According to this view, scientists begin by searching for the truth and examine all evidence with an open mind while on that quest; creationists, however, begin with the supposition that they know what the truth is and then go out and try to find evidence to support their position. Such an approach is the opposite of the scientific method, say most scientists. Schoolchildren should not be told that something unscientific *is* scientific; therefore, creationist ideas should be confined to instruction at home or in churches.

Creationists counter this argument by insisting that their view of the origins of life is just as scientifically based as the one evolutionists promote. At any rate, say creationists, children should be exposed to both views in the classroom. That is the only fair approach because students who are exposed only to evolutionary theory may be deprived of a full hearing on the creationist version. Public schools should not favor one theory over the other, they insist, especially in states, counties, or individual school districts in which a clear majority of parents want their children to learn about the biblical version of creation in school.

One major result of these differing opinions about the teaching of evolution and creationism has been a concerted effort on the part of creationist organizations to pass laws supporting their position. At first these laws sought to ban the teaching of evolution (for example, the 1925 Tennessee Butler Act that sparked the infamous Scopes "Monkey Trial"). When this goal became unrealistic, creationists began trying to get equal time for their views in the classroom alongside evolution.

Up until the close of the twentieth century, in cases in

which such laws were challenged in court, judges tended to side with the pro-evolution side. But that may well change in the future. Recent polls indicate that 60 percent or more of the parents in many areas of the United States favor a balanced presentation of both views of life's origins in high school biology courses. It is possible that some judges will respond to this reality by ruling in favor of the pro-creationist side. If so, the opposing side will surely contest it, for on this issue neither side seems remotely willing to back down or compromise.

Official Tenets of Creationist Belief

INSTITUTE FOR CREATION RESEARCH

The fundamentalist beliefs of modern creationists demonstrate that a number of people continue to reject the theory of evolution and all the biological, geological, and historical implications that go with it. As the creationists see it, Earth is only a few thousand years old, so there has not been enough time for evolution to take place. At the same time, creationists feel that the miraculous acts attributed to God in the Bible are a completely logical and satisfying explanation for how the universe and life originated. Among the many groups and religious sects that advocate creationist doctrines, the Institute for Creation Research (ICR), in Santee, California, is one of the most prominent and well organized. Following is ICR's official statement of the tenets of creationism. It is divided into two broad sections—"Scientific Creationism" and "Biblical Creationism"; however, ICR maintains "that the two are compatible and that all genuine facts of science support the Bible." These tenets of creationism form the core of the creationists' two main arguments, first that evolution is either invalid or merely a theory, and second, that creationism should be taught in public schools alongside evolution.

- The physical universe of space, time, matter, and energy has not always existed, but was supernaturally created by a transcendent personal Creator who alone has existed from eternity.

- The phenomenon of biological life did not develop by natural processes from inanimate systems but was specially and supernaturally created by the Creator.
- Each of the major kinds of plants and animals was created functionally complete from the beginning and did not evolve from some other kind of organism. Changes in basic kinds since their first creation are limited to "horizontal" changes (variation) within the kinds, or "downward" changes (e.g., harmful mutations, extinctions).
- The first human beings did not evolve from an animal ancestry, but were specially created in fully human form from the start. Furthermore, the "spiritual" nature of man (self-image, moral consciousness, abstract reasoning, language, will, religious nature, etc.) is itself a supernaturally created entity distinct from mere biological life.
- The record of earth history, as preserved in the earth's crust, especially in the rocks and fossil deposits, is primarily a record of catastrophic intensities of natural processes, operating largely within uniform natural laws, rather than one of gradualism and relatively uniform process rates. There are many scientific evidences for a relatively recent creation of the earth and the universe, in addition to strong scientific evidence that most of the earth's fossiliferous sedimentary rocks were formed in an even more recent global hydraulic cataclysm [catastrophe].
- Processes today operate primarily within fixed natural laws and relatively uniform process rates but, since these were themselves originally created and are daily maintained by their Creator, there is always the possibility of miraculous intervention in these laws or processes by their Creator. Evidences for such intervention should be scrutinized critically, however, because there must be clear and adequate reason for any such action on the part of the Creator.

- The universe and life have somehow been impaired since the completion of creation, so that imperfections in structure, disease, aging, extinctions, and other such phenomena are the result of "negative" changes in properties and processes occurring in an originally-perfect created order.
- Since the universe and its primary components were created perfect for their purposes in the beginning by a competent and volitional Creator, and since the Creator does remain active in this now-decaying creation, there do exist ultimate purposes and meanings in the universe. Teleological considerations [i.e., considerations of design and purpose in nature], therefore, are appropriate in scientific studies whenever they are consistent with the actual data of observation, and it is reasonable to assume that the creation presently awaits the consummation of the Creator's purpose.
- Although people are finite and scientific data concerning origins are always circumstantial and incomplete, the human mind (if open to the possibility of creation) is able to explore the manifestations of that Creator rationally and scientifically, and to reach an intelligent decision regarding one's place in the Creator's plan.

Tenets of Biblical Creationism

- The Creator of the universe is a triune God-Father, Son, and Holy Spirit. There is only one eternal and transcendent God, the source of all being and meaning, and He exists in three Persons, each of whom participated in the work of creation.
- The Bible, consisting of the thirty-nine canonical books of the Old Testament and the twenty-seven canonical books of the New Testament, is the divinely-inspired revelation of the Creator to man. Its unique, plenary, verbal inspiration guarantees that these writings, as originally and miraculously given, are infallible and completely authoritative on all matters with which

they deal, free from error of any sort, scientific and historical as well as moral and theological.

- All things in the universe were created and made by God in the six literal days of the creation week described in Genesis 1:1–2:3, and confirmed in Exodus 20:8–11. The creation record is factual, historical, and perspicuous; thus all theories of origins or development which involve evolution in any form are false. All things which now exist are sustained and ordered by God's providential care. However, a part of the spiritual creation, Satan and his angels, rebelled against God after the creation and are attempting to thwart His divine purposes in creation.
- The first human beings, Adam and Eve, were specially created by God, and all other men and women are their descendants. In Adam, mankind was instructed to exercise "dominion" over all other created organisms, and over the earth itself (an implicit commission for true science, technology, commerce, fine art, and education) but the temptation by Satan and the entrance of sin brought God's curse on that dominion and on mankind, culminating in death and separation from God as the natural and proper consequence.
- The Biblical record of primeval earth history in Genesis 1–2 is fully historical and perspicuous, including the creation and fall of man, the curse on the creation and its subjection to the bondage of decay, the promised Redeemer, the worldwide cataclysmic deluge in the days of Noah, the post-diluvian renewal of man's commission to subdue the earth (now augmented by the institution of human government) and the origin of nations and languages at the tower of Babel.
- The alienation of man from his Creator because of sin can only be remedied by the Creator Himself, who became man in the person of the Lord Jesus Christ, through miraculous conception and virgin birth. In Christ were indissolubly united perfect sinless hu-

manity and full deity, so that His substitutionary death is the only necessary and sufficient price of man's redemption. That the redemption was completely efficacious is assured by His bodily resurrection from the dead and ascension into heaven; the resurrection of Christ is thus the focal point of history, assuring the consummation of God's purposes in creation.

- The final restoration of creation's perfection is yet future, but individuals can immediately be restored to fellowship with their Creator, on the basis of His redemptive work on their behalf, receiving forgiveness and eternal life solely through personal trust in the Lord Jesus Christ, accepting Him not only as estranged Creator but also as reconciling Redeemer and coming King. Those who reject Him, however, or who neglect to believe on Him, thereby continue in their state of rebellion and must ultimately be consigned to the everlasting fire prepared for the devil and his angels.

Common Creationist Misconceptions About Evolution

MARK ISAAK

In this essay, science writer Mark Isaak explains why he thinks creationist criticisms of the doctrine of evolution are off the mark. Typically, he says, creationists attempt to refute their own misconceptions about evolution rather than the facts about evolution. He refers to such misguided creationist views as the "straw-man version of evolution." For example, Isaak asserts that creationists often claim there are no transitional fossils (remains of creatures whose physical properties show they were in the process of evolving from one species into another) without a clear understanding of what transitional fossils really are; in reality, he says, many transitional fossils have been found.

A large part of the reason why Creationist arguments against evolution can sound so persuasive is because they don't address evolution, but rather argue against a set of misunderstandings that people are right to consider ludicrous. The Creationists wrongly believe that their understanding of evolution is what the theory of evolution really says, and declare evolution banished. In fact, they haven't even addressed the topic of evolution. (The situation isn't helped by poor science education generally. Even

---

most beginning college biology students don't understand the theory of evolution.)

The five propositions below seem to be the most common misconceptions based on a Creationist straw-man version of evolution. If you hear anyone making any of them, chances are excellent that they don't know enough about the real theory of evolution to make informed opinions about it.

- Evolution has never been observed.
- Evolution violates the 2nd law of thermodynamics.
- There are no transitional fossils.
- The theory of evolution says that life originated, and evolution proceeds, by random chance.
- Evolution is only a theory; it hasn't been proved.

Explanations of why these statements are wrong are given below. . . .

Evolution Has Never Been Observed

Biologists define evolution as a change in the gene pool of a population over time. One example is insects developing a resistance to pesticides over the period of a few years. Even most Creationists recognize that evolution at this level is a fact. What they don't appreciate is that this rate of evolution is all that is required to produce the diversity of all living things from a common ancestor. . . .

Even without these direct observations, it would be wrong to say that evolution hasn't been observed. Evidence isn't limited to seeing something happen before your eyes. Evolution makes predictions about what we would expect to see in the fossil record, comparative anatomy, genetic sequences, geographical distribution of species, etc., and these predictions have been verified many times over. The number of observations supporting evolution is overwhelming.

What hasn't been observed is one animal abruptly changing into a radically different one, such as a frog changing into a cow. This is not a problem for evolution because evolution doesn't propose occurrences even remotely like that. In fact, if we ever observed a frog turn into a cow, it would be very strong evidence *against* evolution.

Evolution Violates the Second Law of Thermodynamics

This shows more a misconception about thermodynamics than about evolution. The second law of thermodynamics says, "No process is possible in which the *sole result* is the transfer of energy from a cooler to a hotter body." Now you may be scratching your head wondering what this has to do with evolution. The confusion arises when the 2nd law is phrased in another equivalent way, "The entropy of a closed system cannot decrease." Entropy is an indication of unusable energy and often (but not always!) corresponds to intuitive notions of disorder or randomness. Creationists thus misinterpret the 2nd law to say that things invariably progress from order to disorder.

However, they neglect the fact that life is not a closed system. The sun provides more than enough energy to drive things. If a mature tomato plant can have more usable energy than the seed it grew from, why should anyone expect that the next generation of tomatoes can't have more usable energy still? Creationists sometimes try to get around this by claiming that the information carried by living things lets them create order. However, not only is life irrelevant to the 2nd law, but order from disorder is common in nonliving systems, too. Snowflakes, sand dunes, tornadoes, stalactites, graded river beds, and lightning are just a few examples of order coming from disorder in nature; none require an intelligent program to achieve that order. In any nontrivial system with lots of energy flowing through it, you are almost certain to find order arising somewhere in the system. If order from disorder is supposed to violate the 2nd law of thermodynamics, why is it ubiquitous in nature?

The thermodynamics argument against evolution displays a misconception about evolution as well as about thermodynamics, since a clear understanding of how evolution works should reveal major flaws in the argument. Evolution says that organisms reproduce with only small changes between generations (after their own kind, so to speak). For example, animals might have appendages which are longer or shorter, thicker or flatter, lighter or darker than their parents. Occasionally, a change might be on the order of having four or six fingers instead of five. Once the differences appear, the theory of evolution calls for differential reproductive success. For example, maybe the animals with longer appendages survive to have more offspring than short-appendaged ones. All of these processes can be observed today. They obviously don't violate any physical laws.

There Are No Transitional Fossils

A transitional fossil is one that looks like it's from an organism intermediate between two lineages, meaning it has some characteristics of lineage A, some characteristics of lineage B, and probably some characteristics part way between the two. Transitional fossils can occur between groups of any taxonomic level, such as between species, between orders, etc. Ideally, the transitional fossil should be found stratigraphically between the first occurrence of the ancestral lineage and the first occurrence of the descendent lineage, but evolution also predicts the occurrence of some fossils with transitional morphology [physical shape and structure] that occur after both lineages. There's nothing in the theory of evolution which says an intermediate form (or any organism, for that matter) can have only one line of descendents, or that the intermediate form itself has to go extinct when a line of descendents evolves.

To say there are no transitional fossils is simply false. Paleontology has progressed a bit since *Origin of Species* was published, uncovering thousands of transitional fossils, by both the temporally restrictive and the less restrictive definitions. The fossil record is still spotty and always will be;

erosion and the rarity of conditions favorable to fossilization make that inevitable. Also, transitions may occur in a small population, in a small area, and/or in a relatively short amount of time; when any of these conditions hold, the chances of finding the transitional fossils go down. Still, there are many instances where excellent sequences of transitional fossils exist. Some notable examples are the transitions from reptile to mammal, from land animal to early whale, and from early ape to human. . . .

The misconception about the lack of transitional fossils is perpetuated in part by a common way of thinking about categories. When people think about a category like "dog" or "ant," they often subconsciously believe that there is a well-defined boundary around the category, or that there is some eternal ideal form (for philosophers, the Platonic idea) which defines the category. This kind of thinking leads people to declare that Archaeopteryx is "100% bird," when it is clearly a mix of bird and reptile features (with more reptile than bird features, in fact). In truth, categories are man-made and artificial. Nature is not constrained to follow them, and it doesn't.

Some Creationists claim that the hypothesis of punctuated equilibrium [evolution occurring in short, periodic bursts] was proposed (by [the noted scientists] Eldredge and Gould) to explain gaps in the fossil record. Actually, it was proposed to explain the relative rarity of transitional forms, not their total absence, and to explain why speciation appears to happen relatively quickly in some cases, gradually in others, and not at all during some periods for some species. In no way does it deny that transitional sequences exist. In fact, both Gould and Eldredge are outspoken opponents of Creationism.

> "But paleontologists have discovered several superb examples of intermediary forms and sequences, more than enough to convince any fair-minded skeptic about the reality of life's physical genealogy."—Stephen Jay Gould, *Natural History*, May 1994

The Theory of Evolution Says That Life Originated, and Evolution Proceeds, by Random Chance

There is probably no other statement which is a better indication that the arguer doesn't understand evolution. Chance certainly plays a large part in evolution, but this argument completely ignores the fundamental role of natural selection, and selection is the very opposite of chance. Chance, in the form of mutations, provides genetic variation, which is the raw material that natural selection has to work with. From there, natural selection sorts out certain variations. Those variations which give greater reproductive success to their possessors (and chance ensures that such beneficial mutations will be inevitable) are retained, and less successful variations are weeded out. When the environment changes, or when organisms move to a different environment, different variations are selected, leading eventually to different species. Harmful mutations usually die out quickly, so they don't interfere with the process of beneficial mutations accumulating.

Nor is abiogenesis (the origin of the first life) due purely to chance. Atoms and molecules arrange themselves not purely randomly, but according to their chemical properties. In the case of carbon atoms especially, this means complex molecules are sure to form spontaneously, and these complex molecules can influence each other to create even more complex molecules. Once a molecule forms that is approximately self-replicating, natural selection will guide the formation of ever more efficient replicators. The first self-replicating object didn't need to be as complex as a modern cell or even a strand of DNA. Some self-replicating molecules are not really all that complex (as organic molecules go).

Some people still argue that it is wildly improbable for a given self-replicating molecule to form at a given point (although they usually don't state the "givens," but leave them implicit in their calculations). This is true, but there were oceans of molecules working on the problem, and no one knows how many possible self-replicating molecules could have served as the first one. A calculation of the odds of abiogenesis is worthless unless it recognizes the immense range of starting materials that the first replicator might have formed from, the probably innumerable different forms that the first replicator might have taken, and the fact that much of the construction of the replicating molecule would have been non-random to start with.

(One should also note that the theory of evolution doesn't depend on how the first life began. The truth or falsity of any theory of abiogenesis wouldn't affect evolution in the least.)

Evolution Is Only a Theory; It Hasn't Been Proved

First, we should clarify what "evolution" means. Like so many other words, it has more than one meaning, its strict biological definition is "a change in allele frequencies over time." By that definition, evolution is an indisputable fact. Most people seem to associate the word "evolution" mainly with common descent, the theory that all life arose from one common ancestor. Many people believe that there is enough evidence to call this a fact, too. However, common descent is still not the theory of evolution, but just a fraction of it (and a part of several quite different theories as well). The *theory* of evolution not only says that life evolved, it also includes mechanisms, like mutations, natural selection, and genetic drift, which go a long way towards explaining *how* life evolved.

Calling the theory of evolution "only a theory" is, strictly speaking, true, but the idea it tries to convey is completely wrong. The argument rests on a confusion between what "theory" means in informal usage and in a scientific context. A theory, in the scientific sense, is "a coherent group of general propositions used as principles of explanation for a class of phenomena" [Random House American College Dictionary]. The term does not imply tentativeness or lack of certainty. Generally speaking, scientific theories differ from scientific laws only in that laws can be expressed more tersely. Being a theory implies self-consistency, agreement with observations, and usefulness. (Creationism fails to be a theory mainly because of the last point; it makes few or no specific claims about what we would expect to find, so it can't be used for anything. When it does make falsifiable predictions, they prove to be false.)

Lack of proof isn't a weakness, either. On the contrary, claiming infallibility for one's conclusions is a sign of hubris. Nothing in the real world has ever been rigorously proved, or ever will be. Proof, in the mathematical sense, is possible only if you have the luxury of defining the universe you're operating in. In the real world, we must deal with levels of certainty based on observed evidence. The more and better evidence we have for something, the more certainty we assign to it; when there is enough evidence, we label the something a fact, even though it still isn't 100% certain.

What evolution has is what any good scientific claim has—evidence, and lots of it. Evolution is supported by a wide range of observations throughout the fields of genetics, anatomy, ecology, animal behavior, paleontology, and others. If you wish to challenge the theory of evolution, you must address that evidence. You must show that the evidence is either wrong or irrelevant or that it fits another theory better. Of course, to do this, you must know both the theory and the evidence.

Conclusion

These are not the only misconceptions about evolution by any means. Other common misunderstandings include how geological dating techniques work, implications to morality and religion, the meaning of "uniformitarianism," and many more. To address all these objections here would be impossible.

But consider: About a hundred years ago, scientists, who were then mostly Creationists, looked at the world to figure out how God did things. These Creationists came to the conclusions of an old earth and species originating by evolution. Since then, thousands of scientists have been studying evolution with increasingly more sophisticated tools. Many of these scientists have excellent understandings of the laws of thermodynamics, how fossil finds are interpreted, etc., and finding a better alternative to evolution would win them fame and fortune. Sometimes their work has changed our understanding of significant details of how evolution operates, but the theory of evolution still has essentially unanimous agreement from the people who work on it.

The Scopes Trial Focuses Attention on Teaching Evolution

LYNDSEY McCABE

The most famous confrontation between the supporters and opponents of teaching evolution and/or the biblical explanation of creation in schools was the 1925 Scopes trial, which took place in Tennessee and for twelve days riveted the attention of the nation. As Lyndsey McCabe of the University of Virginia explains in this overview of the event, a local teacher was accused of breaking a new law that prohibited the teaching of evolution in public schools. The trial attracted the participation of the American Civil Liberties Union, as well as two high-profile legal and political figures, the great trial lawyer Clarence Darrow and the conservative orator and frequent candidate for the presidency William Jennings Bryan. McCabe also points out how the so-called "Monkey Trial" ended up setting the tone of the debate over teaching evolution in the classroom for the rest of the twentieth century.

Setting the Stage: The Butler Law

As America emerged from World War I, a collective nostalgia swept the country for the relative simplicity and "normalcy" of prewar society. In rural areas, particularly in the South and Midwest, Americans turned to their faith for comfort and stability, and fundamentalist religion soared

Lyndsey McCabe, "The Scopes 'Monkey Trial'—July 10, 1925–July 25, 1925," http://xroads.virginia.edu, 1996. Copyright © 1996 by the University of Virginia. Reproduced by permission.

in popularity. Fundamentalists, who believed in a literal interpretation of the Bible, locked into Darwin and the theory of evolution as "the most present threat to the truth they were sure they alone possessed." With evolution as the enemy, they set out to eradicate it from their society, beginning with the education system.

By 1925, states across the South had passed laws prohibiting the teaching of evolution in the classroom. Oklahoma, Florida and Mississippi had such laws, and narrow margins determined those in North Carolina and Kentucky. In Tennessee the Butler Law passed in early 1925, for although the governor was not a fundamentalist, many of his constituents were. As he said, "Nobody believes that it is going to be an active statute." No one that is, but the American Civil Liberties Union [ACLU] in New York, which was becoming increasingly more wary of what they saw as an infringement on their constitutional rights. With an eye on Tennessee, the ACLU set out to initiate a court case to test the constitutionality of the Butler Law.

The Curtain Opens on Dayton

Within days of the ACLU's decision to test the Butler Law, George W. Rappelyea spotted a press release in a Tennessee newspaper offering legal support to any teacher who would challenge the law. For Rappelyea, an ardent evolutionist and a Dayton booster, there was no better way to bring down the detested law and promote the small Tennessee mining town. On May 5, Rappelyea and other local leaders met at F.E. Robinson's drug store and hammered out the details of their plan. All they needed was a teacher to test the law, and they found him in John T. Scopes, a 24-year-old science teacher and football coach. When questioned about his teaching of evolution as a part of teaching biology, Scopes replied, "So has every other teacher. Evolution is explained in Hunter's 'Civic Biology,' and that's our textbook."

Scopes was hesitant, at first, to join the case, but Rappelyea was determined. The trial was to be a grand affair and bring fame and fortune to the small town. He began his scheme saying, "Let's take this thing to court and test the legality of it. I will swear out warrant and have you arrested. . . . That will make a big sensation. Why not bring a lot of doctors and preachers here? Let's get H.G. Wells and a lot of big fellows." With Scopes' agreement, Rappelyea wired the ACLU, and "the stage was set . . . the play could open at once."

The Cast and Crew

The Scopes trial met all of Rappelyea's expectations and more. During the twelve hot July days in court, Dayton swarmed with politicians and lawyers, preachers and university scholars, reporters and even circus performers. The streets of Dayton took on the appearance of a small-town fair, with people selling food, souvenirs and religious books. On the side of the courthouse ran a banner blaring "Read Your Bible Daily!" The reporters came from as far away as Hong Kong, and collectively they penned more than two million words during the trial. Chief among the media was H.L. Mencken of the *Baltimore Sun*, known for his caustic wit and cynical observations.

Into this media circus meets religious revival rolled two of the greatest legal minds of the time, facing off to battle each other. William Jennings Bryan called the trial a "contest between evolution and Christianity . . . a duel to the death." Known as The Great Commoner to the people, Bryan was a three-time presidential candidate and former Secretary of State to Woodrow Wilson. After a few years of retirement, he joined the Chautauqua circuit to rail against Darwin in tent revivals across the country.

Across the courtroom at the defendant's table was Clarence Darrow, with a sharp criminal lawyer's mind and an infamous reputation. To Bryan, he was "the greatest atheist or agnostic in the United States." Darrow himself joined the defense table because "for years," he said, "I've wanted to put Bryan in his place as a bigot."

Bryan's Show and Darrow's Finale

From the moment of Bryan's arrival in Dayton, the weight of public sentiment was in his favor. The records of the trial indicate that the townspeople came out for the trial in record numbers, packing the small country courthouse. Cries of "Amen" peppered the trial proceedings until the judge had to ask the observers to lower the noise level. Bryan planned to end the trial with a speech consummating his lifetime of preaching, one he had been preparing for seven weeks. Darrow, however, had other plans. Since the intention was to test the constitutionality of the Butler Law, Darrow wanted the jury to find Scopes guilty, so he could then appeal the decision in a higher court. He did not, however, plan to call Scopes to the stand, for if he were to do so, it might surface that Scopes had, in fact, not even been in school on the day mentioned in the indictment. He was meticulous in his effort to keep the trial free of technicalities. Just one could get the case thrown out with the law itself yet untested. Darrow also planned to call expert witnesses to give testimony about evolution. But when the judge ordered that Darrow could not call the scholars as witnesses, he shifted his plans.

After the judge moved the trial outside because of the 100-plus degree heat inside and the instability of the courtroom floor under the weight of so many spectators, Darrow, in a fantastic gesture, called William Jennings Bryan to the stand. The interchange which follows targets the essence of Darrow's argument and signals the turning point in the trial, which brought public sentiment decisively over to Darrow's side:

> "You have given considerable study to the Bible, haven't you, Mr. Bryan?"
>
> "Yes, sir; I have tried to. . . . But, of course, I have studied it more as I have become older than when I was a boy."
>
> "Do you claim then that everything in the Bible should be literally interpreted?"

"I believe everything in the Bible should be accepted as it is given there. . . ."

Darrow continued to question Bryan on the actuality of Jonah and the whale, Joshua's making the sun stand still and the Tower of Babel, as Bryan began to have more difficulty answering.

Q: "Do you think the earth was made in six days?"

A: "Not six days of 24 hours. . . . My impression is they were periods. . . ."

Q: "Now, if you call those periods, they may have been a very long time?"

A: "They might have been."

Q: "The creation might have been going on for a very long time?"

A: "It might have continued for millions of years. . . ."

Darrow had set his trap and Bryan walked right in. Darrow asked for and was granted an immediate direct verdict, thereby blocking Bryan from giving his speech. Within eight minutes of deliberation, the jury returned with a verdict of guilty and the judge ordered Scopes to pay a fine of $100, the minimum the law allowed; In his last words to the court, Scopes, the man who was reluctant from the start, said, "Your Honor, I feel that I have been convicted of violating an unjust statute. I will continue in the future . . . to oppose this law in any way I can. Any other action would be in violation of my idea of academic freedom."

Just five days after the trial ended, Bryan lay down for a Sunday afternoon nap and never woke up. The diabetes with which he had contended for years had finally taken his life.

The trial itself also passed on when more than a year later, on January 14, 1927, the State Supreme Court in Nashville handed down a decision which reversed the earlier one. However, the court's decision stemmed from the very point Darrow sought to avoid—a technicality. By Tennessee state law, the jury, not the judge, must set the fine if it is above $50. The Butler Law, then, stood untested.

Scopes' Place in Culture

The Scopes trial came at a crossroads in history—as people were choosing to cling to the past or jump into the future. The trial itself was a series of conflicts, the obvious one being evolution vs. religion. But . . . there were a series of tensions throughout the trial, including questions of collective vs. individual rights and academic vs. parental concerns, which have persisted in American culture since the birth of the nation. At issue in both of these conflicts was who had control of the society. Who controlled the schools—the masses or the teachers? Who determined the law—the people or the leaders of the town? The resolution was even more unsettling because there was none. Scopes lost the case, but won the public's favor, and the Butler Law remained on the books in Tennessee.

For historical scholars, understanding the Scopes trial begins with a cultural framework. . . . As R.N.Cornelius wrote in "Their Stage Drew All the World," "This controversy, whose stage was the battle over the nature of the Bible, produced a whole cycle of dramatic confrontations, of which the Scopes trial was but one."

The Scopes trial was not distinct, therefore so much for its theme as it was for its presentation. Other school districts and other towns, struggling with this very issue, missed the media circus. Dayton, however, came to center-stage, with the lawyers and business men writing the script and the country enthralled with this true American drama.

Evolution vs. Creationism in the Classroom: The Core Issues

PARTICIPANTS IN THE SCOPES TRIAL

The Scopes trial not only focused national attention on the growing confrontation between science and religion in the United States, but also sharply delineated most of the core issues involved in the more specific debate about whether evolution or creationism (or both) should be taught in American classrooms. Many parts of the trial transcripts address these issues, including the two excerpts reproduced here. In the first, Howard Morgan, a student whom Scopes had taught the basics of evolution, is questioned first by prosecutor A. Thomas Stewart and then by defense counsel Clarence Darrow. The second excerpt from the transcripts consists of the closing speech that chief prosecutor William Jennings Bryan never got to deliver (because the defense asked the judge to return a verdict of "guilty" so that the case could be appealed to the state supreme court). A powerful statement, the speech draws the battle lines between science and evolution and religion and creationism, emphasizing in particular the creationist reliance on faith in Jesus Christ and the "bloody, brutal doctrine" of evolution.

Howard Morgan, testimony, *Tennessee v. John Scopes*, 1925, and William Jennings Bryan, closing speech, *Tennessee v. John Scopes*, 1925.

Direct Examination of Howard Morgan by General Stewart:

Q—Your name is Howard Morgan?

A—Yes, sir.

Q—You are Mr. Luke Morgan's son?

A—Yes, sir. . . .

Q—Your Father is in the bank here, Dayton Bank and Trust company?

A—Yes, sir.

Q—How old are you?

A—14 years.

Q—Did you attend school here at Dayton last year?

A—Yes, sir.

Q—What school?

A—High School.

Q—Central High School?

A—Yes, sir.

Q—Did you study anything under Prof. Scopes?

A—Yes, sir.

Q—Did you study this book, General Science?

A—Yes, sir. . . .

Q—Were you studying that book in April of this year, Howard?

A—Yes, sir.

Q—Did Prof. Scopes teach it to you?

A—Yes, sir.

Q—When did you complete the book?

A—Latter part of April.

Q—When was school out?

A—First or second of May.

Q—You studied it then up to a week or so before school was out?

A—Yes, sir.

Q—Now, you say you were studying this book in April; how did Prof. Scopes teach that book to you? I mean by that did he ask you questions and you answered them or did he give you lectures, or both? Just explain to the jury here now, these gentlemen here in front of you, how he taught the books to you.

A—Well, sometimes he would ask us questions and then he would lecture to us on different subjects in the book.

Q—Sometimes he asked you questions and sometimes lectured to you on different subjects in the book?

A—Yes, sir.

Q—Did he ever undertake to teach you anything about evolution?

A—Yes, sir. . . .

Q—Just state in your own words, Howard, what he taught you and when it was.

A—It was along about the 2d of April.

Q—Of this year?

A—Yes, sir; of this year. He said that the earth was once a hot molten mass too hot for plant or animal life to exist upon it; in the sea the earth cooled off; there was a little germ of one cell organism formed, and this organism kept evolving until it got to be a pretty good-sized animal, and then came on to be a land animal and it kept on evolving, and from this was man.

Q—Let me repeat that; perhaps a little stronger than you. If I don't get it right you correct me.

Hays—Go to the head of the class. . . .

Stewart—I ask you further, Howard, how did he classify man with reference to other animals; what did he say about them?

A—Well, the book and he both classified man along with cats and dogs, cows, horses, monkeys, lions, horses and all that.

Q—What did he say they were?

A—Mammals.

Q—Classified them along with dogs, cats, horses, monkeys and cows?

A—Yes, sir.

Cross Examination by Mr. Darrow:

Q—Let's see, your name is what?

A—Howard Morgan.

Q—Now, Howard, what do you mean by classify?

A—Well, it means classify these animals we mentioned, that men were just the same as them, in other words—

Q—He didn't say a cat was the same as a man?

A—No, sir: he said man had a reasoning power; that these animals did not.

Q—There is some doubt about that, but that is what he said, is it? (Laughter in the courtroom.)

The Court—Order.

Stewart—With some men.

Darrow—A great many.

Q—Now, Howard, he said they were all mammals, didn't he?

A—Yes, sir.

Q—Did he tell you what a mammal was, or don't you remember?

A—Well, he just said these animals were mammals and man was a mammal.

Q—No; but did he tell you what distinguished mammals from other animals?

A—I don't remember.

Q—If he did, you have forgotten it? Didn't he say that mammals were those beings which suckled their young?

A—I don't remember about that.

Q—You don't remember?

A—No.

Q—Do you remember what he said that made any animal a mammal, what it was or don't you remember?

A—I don't remember.

Q—But he said that all of them were mammals?

A—All what?

Q—Dogs and horses, monkeys, cows, man, whales, I cannot state all of them, but he said all of those were mammals?

A—Yes, sir; but I don't know about the whales; he said all those other ones. (Laughter in the courtroom.)

The Court—Order. . . .

Q—Well, did he tell you anything else that was wicked?

A—No, not that I remember of. . . .

Q—Now, he said the earth was once a molten mass of liquid, didn't he?

A—Yes.

Q—By molten, you understand melted?

A—Yes, sir.

Q—After that, it got cooled enough and the soil came, that plants grew; is that right?

A—Yes, sir, yes, sir.

Q—And that the first life was in the sea.

Q—And that it developed into life on the land?

A—Yes, sir.

Q—And finally into the highest organism which is known to man?

A—Yes, sir.

Q—Now, that is about what he taught you?

Q—It has not hurt you any, has it?

A—No, sir.

Darrow—That's all.

Bryan's Closing Speech

Science is a magnificent force, but it is not a teacher of morals. It can perfect machinery, but it adds no moral restraints to protect society from the misuse of the machine. It can also build gigantic intellectual ships, but it constructs no moral rudders for the control of storm tossed human vessels. It not only fails to supply the spiritual element needed but some of its unproven hypotheses rob the ship of its compass and thus endangers its cargo. In war, science has proven itself an evil genius; it has made war more terrible than it ever was before. Man used to be content to slaughter his fellowmen on a single plane—the earth's surface. Science has taught him to go down into the water and shoot up from below and to go up into the clouds and shoot down from above, thus making the battlefield three times as bloody as it was before; but science does not teach brotherly love. Science has made war so hellish that civilization was about to commit suicide; and now we are told that newly discovered instruments of destruction will make the cruelties of the late war seem trivial in comparison with the cruelties of wars that may come in the future. If civilization is to be saved from the wreckage threatened by intelligence not consecrated by love, it must be saved by the moral code of the meek and lowly Nazarene [i.e., Jesus Christ] His teachings, and His teachings, alone, can solve the problems that vex heart and perplex the world. . . .

It is for the jury to determine whether this attack upon the Christian religion shall be permitted in the public schools of Tennessee by teachers employed by the state and paid out of the public treasury. This case is no longer local, the defendant ceases to play an important part. The case has assumed the proportions of a battle-royal between unbelief that attempts to speak through so-called science and the defenders of the Christian faith, speaking through the legislators of Tennessee. It is again a choice between God and Baal [an ancient pagan god]; it is also a renewal of the issue in Pilate's court. . . .

Again force and love meet face to face, and the question, "What shall I do with Jesus?" must be answered. A bloody, brutal doctrine—Evolution—demands, as the rabble did nineteen hundred years ago, that He be crucified. That cannot be the answer of this jury representing a Christian state and sworn to uphold the laws of Tennessee. Your answer will be heard throughout the world; it is eagerly awaited by a praying multitude. If the law is nullified, there will be rejoice wherever God is repudiated, the savior scoffed at and the Bible ridiculed. Every unbeliever of every kind and degree will be happy. If, on the other hand, the law is upheld and the religion of the school children protected, millions of Christians will call you blessed and, with hearts full of gratitude to God, will sing again that grand old song of triumph: "Faith of our fathers, living still, In spite of dungeon, fire and sword; O how our hearts beat high with joy Whenever we hear that glorious word—Faith of our fathers—Holy faith; We will be true to thee till death!"

Creationism Should Be Taught in Public Schools

KERWIN THIESSEN

Creationists have advanced a number of arguments in favor of teaching their version of the origins of life in schools. The main ones, including the assertion that both evolution and creationism are belief systems, are neatly summarized in this well-crafted essay by Kerwin Thiessen, pastor of the Koerner Heights Mennonite Brethren Church in Newton, Kansas.

S everal years ago the *Wichita Eagle-Beacon* carried a story which reported the Flat Earth Society to be alive and well in modern America. This select group of traditionalists have held tenaciously to the archaic notion that the earth is flat, not spherical. Such a society reminds one of the self-proclaimed scientist who said, "Don't confuse me with the facts. My mind is made up!"

In the most recent court battle concerning this issue, U.S. District Judge William Overton struck down Arkansas' "Balanced Treatment for Creation-Science and Evolution-Science Act" by rejecting two state arguments that are crucial elements in this debate: 1) that evolution is as religious as creationism, and 2) that public schools should teach what the public wants. In striking down the Arkansas law requiring schools to give a balanced presentation of both evolution and creationism Judge Overton stated, "No group, no matter how large or small, may use the organs of movement, of which the public schools are the most conspicuous and influential, to foist its religious beliefs on others."

Is Evolution a Kind of Religion?

What about the question, "Is evolution just as religious as creationism?" My answer is, "yes," because the system of evolution is, in its definition, a belief system. Sir Julian Huxley, one of today's most respected evolutionists has defined evolution as

> a directional and essentially irreversible process occurring in time, which in its course gives rise to an increase of variety and an increasingly high level of organization in its products. Our present knowledge indeed forces us to the view that the whole of reality is evolution—a single process of self-transformation (*What Is Science?* p. 272).

I refer to such a definition because there is not a shred of scientific evidence to scientifically demonstrate Huxley's statement. The definition of *science* shows that something is scientific only if it can be observed and verified. All must admit that it is impossible to prove scientifically any particular concept of origins, creationism or evolution. No human being was there in the beginning to observe and verify how matter and life came into existence.

For nearly one hundred years evolution-as-a-dogma has been accepted and perpetuated in America's public educational institutions, and the proof of such a statement is seen in the almost universal acceptance of three sub-beliefs: 1) faith in spontaneous generation of life substance, 2) faith in transitional forms between different kinds of organisms (unverifiable on the basis of anatomy, embryology, blood and protein analyses, fossils, or genetics), and 3) faith in mutations as a source of raw materials by which supposed evolutionary changes in organisms might have come about in the past. Each of these three sub-beliefs is dependent upon the other, as well as the very definition of *evolution* as given by Huxley. And each of these statements is made without experimentally verifiable or observable scientific evidence.

The point is that both evolution and creationism are belief systems. Both are embraced through "believing what is unobservable" rather than on the basis of what is scientifically provable. I disagree with Judge Overton in striking down the Arkansas law on this basis. He has closed his mind to the true definitions of *science* and *religion. Science* deals with observable data. *Religion* deals with belief and faith in the unobservable.

Give the Public What It Wants

What about the matter of public education, i.e., "Should the public schools teach what the public wants?" The First Amendment of the United States Constitution, Section One, states,

> Congress shall make no law respecting an establishment of religion, or prohibiting the free exercise thereof; or abridging the freedom of speech, or of the press; or the right of the people peaceably to assemble, and to petition the Government for a redress of grievances.

Clearly the intent of our forefathers was that freedom prevail in all areas of civil life, education included. Neither creationism nor evolution should be barred from the classroom where citizens of the republic desire that they be taught. Should not public education be nonpartisan? Some students and their parents will place their faith in the belief-system of evolution when it comes to the study of origins. Others will embrace creationism. Both should be presented in the classroom *where both are desired by the public.*

The role of today's educator should permit such an approach. The teacher used to be seen as a disseminator of knowledge, imparting to eager students the information and concepts they should know. Today's teacher, however, is taught to be skilled in questioning techniques, particularly in the disciplines of science and social science instruction.

This *inquiry approach* puts priority upon a presentation of data by the teacher coupled with thorough reading and experimentation by the student. Such an approach is intended to be objective with the result of developing logical thought and decision-making skills among the students rather than reflecting the pre-conceived bias of the teacher. Such skills as observation, classification, inferring, predicting, measuring, communicating, interpreting, formulating questions and hypothesis, experimentation, and formulating models, teach students to make critical observations when dealing with data.

It makes "First Amendment" sense to present students with all available data in their "search for truth" in any given area of study. Especially in a free society, students must be allowed the academic freedom to pursue every viewpoint possible. The most valid conclusion is always the one that best reflects *all* available data. One certainly cannot label our public schools as "free-learning-centers" when a most crucial area of study, i.e., origins, is oppressed by state and federal legislation permitting only one viewpoint.

Two Competing Models Compared

My answer to the question, "Should creationism be taught in public schools?", is, therefore, "Yes! It should be allowed to be taught in those local school districts where students and parents embrace creationism and desire that it be taught by knowledgeable instructors." My specific rationale for answering in this way is as follows:

1. The constituents of each local school district should determine the parameters of educational instruction for that district and that district only.

2. Since neither evolution nor creation is accessible to the scientific method, since they deal with origins, not presently observable events, both should be formulated as scientific models, or frameworks, within which the student can then predict and correlate observed facts.

3. Such scientific models can neither be proven nor tested, only compared. Which model can explain, with the least amount of difficulty, the observable data in the real world?

4. Evolution is a faith-system. Creation is also a faith-system. Both of these systems have far-reaching implications for the life, world-view, and morality of the adherent. Both systems should be discussed in their entirety and presented as viable options. Therefore, each is as religious as the other. Each is as scientific as the other, as well.

5. Neither evolution nor creationism should be the only view presented in the public classroom if both views are desired by parents and students. Henry Morris of Institute for Creation Research has stated, and I agree, "There are ... strong scientific and pedagogical reasons why *both* models should be taught, as objectively as possible, in public classrooms, giving arguments pro and con for each. Some students and their parents believe in creation, some in evolution, and some are undecided. If creationists desire *only* the creation model to be taught, they should send their children to private schools which do this; if evolutionists want only evolution to be taught, they should provide private schools for *that* purpose. The public schools should be neutral and either teach both or teach neither" (*ICR Impact*, Vol. 1, p. 1).

Teaching Creationism in the Classroom Is Misguided

NATIONAL ACADEMY OF SCIENCES

In this concise, well-worded statement, members of the National Academy of Sciences, one of the most prestigious scientific organizations in the world, make the case that teaching the doctrine of divine creation in the classroom as though it is a legitimate competing theory with evolution is misguided. The assertions made by creationists, the academy spokespersons insist, do not reflect the large body of scientific discoveries that have been made in recent centuries.

The term "evolution" usually refers to the biological evolution of living things. But the processes by which planets, stars, galaxies, and the universe form and change over time are also types of "evolution." In all of these cases there is change over time, although the processes involved are quite different.

In the late 1920s the American astronomer Edwin Hubble made a very interesting and important discovery. Hubble made observations that he interpreted as showing that distant stars and galaxies are receding from Earth in every direction. Moreover, the velocities of recession increase in

National Academy of Sciences, "The Origin of the Universe, Earth, and Life," *Science and Creationism: A View from the National Academy of Sciences*. Washington, DC: National Academy Press, 1999. Copyright © 1999 by the National Academy of Sciences, courtesy of the National Academies Press, Washington, DC. All rights reserved. Reproduced by permission.

proportion with distance, a discovery that has been confirmed by numerous and repeated measurements since Hubble's time. The implication of these findings is that the universe is expanding.

Hubble's hypothesis of an expanding universe leads to certain deductions. One is that the universe was more condensed at a previous time. From this deduction came the suggestion that all the currently observed matter and energy in the universe were initially condensed in a very small and infinitely hot mass. A huge explosion, known as the Big Bang, then sent matter and energy expanding in all directions.

This Big Bang hypothesis led to more testable deductions. One such deduction was that the temperature in deep space today should be several degrees above absolute zero. Observations showed this deduction to be correct. In fact, the Cosmic Microwave Background Explorer (COBE) satellite launched in 1991 confirmed that the background radiation field has exactly the spectrum predicted by a Big Bang origin for the universe.

The Expanding Universe

As the universe expanded, according to current scientific understanding, matter collected into clouds that began to condense and rotate, forming the forerunners of galaxies. Within galaxies, including our own Milky Way galaxy, changes in pressure caused gas and dust to form distinct clouds. In some of these clouds, where there was sufficient mass and the right forces, gravitational attraction caused the cloud to collapse. If the mass of material in the cloud was sufficiently compressed, nuclear reactions began and a star was born.

Some proportion of stars, including our sun, formed in the middle of a flattened spinning disk of material. In the case of our sun, the gas and dust within this disk collided and aggregated into small grains, and the grains formed into larger bodies called planetesimals ("very small planets"), some of which reached diameters of several hundred kilometers. In successive stages these planetesimals coalesced into the nine planets and their numerous satellites. The rocky planets, including Earth, were near the sun, and the gaseous planets were in more distant orbits.

The ages of the universe, our galaxy, the solar system, and Earth can be estimated using modern scientific methods. The age of the universe can be derived from the observed relationship between the velocities of and the distances separating the galaxies. The velocities of distant galaxies can be measured very accurately, but the measurement of distances is more uncertain. Over the past few decades, measurements of the Hubble expansion have led to estimated ages for the universe of between 7 billion and 20 billion years, with the most recent and best measurements within the range of 10 billion to 15 billion years.

The age of the Milky Way galaxy has been calculated in two ways. One involves studying the observed stages of evolution of different-sized stars in globular clusters. Globular clusters occur in a faint halo surrounding the center of the Galaxy, with each cluster containing from a hundred thousand to a million stars. The very low amounts of elements heavier than hydrogen and helium in these stars indicate that they must have formed early in the history of the Galaxy, before large amounts of heavy elements were created inside the initial generations of stars and later distributed into the interstellar medium through supernova explosions (the Big Bang itself created primarily hydrogen and helium atoms). Estimates of the ages of the stars in globular clusters fall within the range of 11 billion to 16 billion years.

Radioactive Elements Tell a Tale

A second method for estimating the age of our galaxy is based on the present abundances of several long-lived radioactive elements in the solar system. Their abundances are set by their rates of production and distribution through exploding supernovas. According to these calculations, the age of our galaxy is between 9 billion and 16 billion years. Thus, both ways of estimating the age of the

Milky Way galaxy agree with each other, and they also are consistent with the independently derived estimate for the age of the universe.

Radioactive elements occurring naturally in rocks and minerals also provide a means of estimating the age of the solar system and Earth. Several of these elements decay with half lives between 700 million and more than 100 billion years (the half life of an element is the time it takes for half of the element to decay radioactively into another element). Using these time-keepers, it is calculated that meteorites, which are fragments of asteroids, formed between 4.53 billion and 4.58 billion years ago (asteroids are small "planetoids" that revolve around the sun and are remnants of the solar nebula that gave rise to the sun and planets). The same radioactive time-keepers applied to the three oldest lunar samples returned to Earth by the Apollo astronauts yield ages between 4.4 billion and 4.5 billion years, providing minimum estimates for the time since the formation of the moon.

The oldest known rocks on Earth occur in northwestern Canada (3.96 billion years), but well-studied rocks nearly as old are also found in other parts of the world. In Western Australia, zircon crystals encased within younger rocks have ages as old as 4.3 billion years, making these tiny crystals the oldest materials so far found on Earth.

The best estimates of Earth's age are obtained by calculating the time required for development of the observed lead isotopes in Earth's oldest lead ores. These estimates yield 4.54 billion years as the age of Earth and of meteorites, and hence of the solar system.

Molecules and the Origins of Life

The origins of life cannot be dated as precisely, but there is evidence that bacteria-like organisms lived on Earth 3.5 billion years ago, and they may have existed even earlier, when the first solid crust formed, almost 4 billion years ago. These early organisms must have been simpler than the organisms living today. Furthermore, before the earliest organisms there must have been structures that one would not call "alive" but that are now components of living things. Today, all living organisms store and transmit hereditary information using two kinds of molecules: DNA and RNA. Each of these molecules is in turn composed of four kinds of subunits known as nucleotides. The sequences of nucleotides in particular lengths of DNA or RNA, known as genes, direct the construction of molecules known as proteins, which in turn catalyze biochemical reactions, provide structural components for organisms, and perform many of the other functions on which life depends. Proteins consist of chains of subunits known as amino acids. The sequence of nucleotides in DNA and RNA therefore determines the sequence of amino acids in proteins; this is a central mechanism in all of biology.

Experiments conducted under conditions intended to resemble those present on primitive Earth have resulted in the production of some of the chemical components of proteins, DNA, and RNA. Some of these molecules also have been detected in meteorites from outer space and in interstellar space by astronomers using radio-telescopes. Scientists have concluded that the "building blocks of life" could have been available early in Earth's history.

An important new research avenue has opened with the discovery that certain molecules made of RNA, called ribozymes, can act as catalysts in modern cells. It previously had been thought that only proteins could serve as the catalysts required to carry out specific biochemical functions. Thus, in the early prebiotic world, RNA molecules could have been "autocatalytic"—that is, they could have replicated themselves well before there were any protein catalysts (called enzymes). Laboratory experiments demonstrate that replicating autocatalytic RNA molecules undergo spontaneous changes and that the variants of RNA molecules with the greatest autocatalytic activity come to prevail in their environments. Some scientists favor the hypothesis that there was an early "RNA world," and they are testing models that lead from RNA to the synthesis of simple DNA and protein molecules. These assemblages of molecules eventually could have become packaged within membranes, thus making up "protocells"—early versions of very simple cells.

For those who are studying the origin of life, the question is no longer whether life could have originated by chemical processes involving nonbiological components. The question instead has become which of many pathways might have been followed to produce the first cells.

Will we ever be able to identify the path of chemical evolution that succeeded in initiating life on Earth? Scientists are designing experiments and speculating about how early Earth could have provided a hospitable site for the segregation of molecules in units that might have been the first living systems. The recent speculation includes the possibility that the first living cells might have arisen on Mars, seeding Earth via the many meteorites that are known to travel from Mars to our planet.

Of course, even if a living cell were to be made in the laboratory, it would not prove that nature followed the same pathway billions of years ago. But it is the job of science to provide plausible natural explanations for natural phenomena. The study of the origin of life is a very active research area in which important progress is being made, although the consensus among scientists is that none of the current hypotheses has thus far been confirmed. The history of science shows that seemingly intractable problems like this one may become amenable to solution later, as a result of advances in theory, instrumentation, or the discovery of new facts.

The Creationist Viewpoint
Many religious persons, including many scientists, hold that God created the universe and the various processes driving physical and biological evolution and that these

processes then resulted in the creation of galaxies, our solar system, and life on Earth. This belief, which sometimes is termed "theistic evolution," is not in disagreement with scientific explanations of evolution. Indeed, it reflects the remarkable and inspiring character of the physical universe revealed by cosmology, paleontology molecular biology, and many other scientific disciplines.

The advocates of "creation science" hold a variety of viewpoints. Some claim that Earth and the universe are relatively young, perhaps only 6,000 to 10,000 years old. These individuals often believe that the present physical form of Earth can be explained by "catastrophism," including a worldwide flood, and that all living things (including humans) were created miraculously, essentially in the forms we now find them.

Other advocates of creation science are willing to accept that Earth, the planets, and the stars may have existed for millions of years. But they argue that the various types of organisms, and especially humans, could only have come about with supernatural intervention, because they show "intelligent design."

In this booklet, both these "Young Earth" and "Old Earth" views are referred to as "creationism" or "special creation."

There are no valid scientific data or calculations to substantiate the belief that Earth was created just a few thousand years ago. This document has summarized the vast amount of evidence for the great age of the universe, our galaxy, the solar system, and Earth from astronomy, astrophysics, nuclear physics, geology, geochemistry, and geophysics. Independent scientific methods consistently give an age for Earth and the solar system of about 5 billion years, and an age for our galaxy and the universe that is two to three times greater. These conclusions make the origin of the universe as a whole intelligible, lend coherence to many different branches of science, and form the core conclusions of a remarkable body of knowledge about the origins and behavior of the physical world.

Nor is there any evidence that the entire geological record, with its orderly succession of fossils, is the product of a single universal flood that occurred a few thousand years ago, lasted a little longer than a year, and covered the highest mountains to a depth of several meters. On the contrary, intertidal and terrestrial deposits demonstrate that at no recorded time in the past has the entire planet been under water. Moreover, a universal flood of sufficient magnitude to form the sedimentary rocks seen today, which together are many kilometers thick, would require a volume of water far greater than has ever existed on and in Earth, at least since the formation of the first known solid crust about 4 billion years ago. The belief that Earth's sediments, with their fossils, were deposited in an orderly sequence in a year's time defies all geological observations and physical principles concerning sedimentation rates and possible quantities of suspended solid matter.

Geologists have constructed a detailed history of sediment deposition that links particular bodies of rock in the crust of Earth to particular environments and processes. If petroleum geologists could find more oil and gas by interpreting the record of sedimentary rocks as having resulted from a single flood, they would certainly favor the idea of such a flood, but they do not. Instead, these practical workers agree with academic geologists about the nature of depositional environments and geological time. Petroleum geologists have been pioneers in the recognition of fossil deposits that were formed over millions of years in such environments as meandering rivers, deltas, sandy barrier beaches, and coral reefs.

The example of petroleum geology demonstrates one of the great strengths of science. By using knowledge of the natural world to predict the consequences of our actions, science makes it possible to solve problems and create opportunities using technology. The detailed knowledge required to sustain our civilization could only have been derived through scientific investigation.

Creationism Is Not Science

The arguments of creationists are not driven by evidence that can be observed in the natural world. Special creation or supernatural intervention is not subjectable to meaningful tests, which require predicting plausible results and then checking these results through observation and experimentation. Indeed, claims of "special creation" reverse the scientific process. The explanation is seen is unalterable, and evidence is sought only to support a particular conclusion by whatever means possible. . . .

Science is not the only way of acquiring knowledge about ourselves and the world around us. Humans gain understanding in many other ways, such as through literature, the arts, philosophical reflection, and religious experience. Scientific knowledge may enrich aesthetic and moral perceptions, but these subjects extend beyond science's realm, which is to obtain a better understanding of the natural world.

The claim that equity demands balanced treatment of evolutionary theory and special creation in science classrooms reflects a misunderstanding of what science is and how it is conducted. Scientific investigators seek to understand natural phenomena by observation and experimentation. Scientific interpretations of facts and the explanations that account for them therefore must be testable by observation and experimentation.

Creationism, intelligent design, and other claims of supernatural intervention in the origin of life or of species are not science because they are not testable by the methods of science. These claims subordinate observed data to statements based on authority, revelation, or religious belief. Documentation offered in support of these claims is typically limited to the special publications of their advocates. These publications do not offer hypotheses subject to change in light of new data, new interpretations, or demonstration of error. This contrasts with science, where any hypothesis or theory always remains subject to the possibility of rejection or modification in the light of new knowledge.

No body of beliefs that has its origin in doctrinal material rather than scientific observation, interpretation, and experimentation should be admissible as science in any science course. Incorporating the teaching of such doctrines into a science curriculum compromises the objectives of public education. Science has been greatly successful at explaining natural processes, and this has led not only to increased understanding of the universe but also to major improvements in technology and public health and welfare. The growing role that science plays in modern life requires that science, and not religion, be taught in science classes.

Modern Developments in and Challenges to Evolution

Chapter Preface

Science has made extraordinary progress since Charles Darwin published his book on the origin of species in 1859. In particular, a host of new discoveries and theories in molecular biology and genetics have significantly expanded and in some cases strongly qualified Darwin's vision of evolution. He did not know about genes and DNA, for example. In fact, the entire field of genetics, which is essential to understanding the structure of living things, did not develop until the early 1900s, well after Darwin's death.

The question is how Darwin's original theory of natural selection—in his view the driving force behind evolution—met the challenge of these later discoveries and developments. Did they strengthen his case or weaken it? Can the newer ideas be reconciled with the older ones? The answer is that Darwin's original theory has survived, but it has undergone, and continues to undergo, reevaluation and alteration as new data reshapes scientists' views of evolution. Noted anthropologist and popular author Stephen Jay Gould put it this way:

> I do believe that the Darwinian framework [for evolution] . . . persists in the emerging structure of a more adequate [modern] evolutionary theory. But I also hold . . . that substantial changes, introduced during the . . . twentieth century, have built a structure so expanded beyond the original explanation that the full exposition, while remaining within the domain of Darwinian logic, must be construed as basically different from the . . . theory of natural selection, rather than simply extended.

To make these developments more understandable, Gould and other scientists sometimes use an architectural analogy. In a way, they say, post-Darwinian evolutionary theory is like the series of additions made to the famous Duomo in Milan, Italy. The original cathedral rose in medieval times in the Gothic style of architecture. In the sixteenth century, sections were added in the Classical Baroque style; other features were added later, including rows of new spires at the top in the nineteenth century. Yet though the structure displays a mixture of styles that makes some architectural purists grumble, most people agree that the original building has not been ruined and the overall effect is impressive. In a similar manner, Gould suggests,

> the modern versions [of evolution] accept the validity of the central logic [of Darwin's natural selection] as a foundation, and introduce their critiques as helpful auxiliaries or additions that enrich, or substantially alter, the original Darwinian formulation, but that leave the kernel of natural selection intact. Thus, the modern reformulations [of evolution] are helpful rather than destructive.

Thus, the discovery of DNA, the molecule that carries the blueprints for life, inevitably changed scientists' view of how natural selection works. Yet it did not disprove natural selection. In fact, few scientists are willing to dispose of Darwin completely, as the case of biochemist Michael J. Behe illustrates. He created much controversy in the late 1990s when he proposed that natural selection may not function at all on the molecular level (where he suspects that something more miraculous may be at work); yet he believes that natural selection is a major and measurable factor on the larger level of animal organs and populations. In this way, progress in evolutionary theory has distanced itself considerably from the original ideas of Charles Darwin yet continues to rely on them as an underlying foundation.

Current Evolutionary Theory

ROBERT WESSON

*Scientists continued to develop and refine the theory of evo-
lution after the pioneering work of Darwin and Wallace.
Robert Wesson, a research fellow at the Hoover Institute in
Stanford, California, here offers a well-informed overview of
advances in evolutionary theory in the twentieth century. As
he points out, though some researchers differ on the exact
manner in which evolution occurs, most scientists accept that
it does somehow occur.*

A generation before Darwin, Jean-Baptiste Lamarck,
who coined the word *biology*, related fossils to living
organisms and proposed a consistent theory of evolution
to explain their differences. Charles's grandfather Erasmus
and others had put forward the idea that species grew out
of one another. Several writers in the first part of the nineteenth century also proposed something like natural selection of those more capable of self-propagation. But Darwin crystallized these ideas in the intellectual atmosphere
and mustered a mass of evidence in their support. No one
else treated the matter with comparable learning and per-
suasiveness and Darwin's passion for detail appealed to ex-
perimental scientists.

Darwin's claim to greatness rests not on his advocacy of
the idea of common origins of species but on his simple
and logical explanation by inheritable variation and natural

selection. This idea was by no means as clear, however, as the thesis of evolution ("We are descended from apes"), and it raised many questions. It was not thoroughly accepted until much later when it was fortified with Mendelian genetics [the theories of Austrian botanist Gregor Mendel] and, in [the twentieth] century, formed the neo-Darwinist synthesis regnant today.

Without Darwin, the doctrine of evolution would probably have been named Wallaceism, after Alfred Russell Wallace. Wallace came to conclusions very close to Darwin's about the same time but received much less credit, partly because he did not support his ideas with such a mass of observations. . . .

Darwin—who was better situated, presented more evidence, and was more consistent in his scientific attitude—became the symbol of evolution personified. Acceptance or denial of the theory of evolution came to be and has remained nearly equivalent to loyalty or opposition to Darwin. Nonetheless, his theory of change by natural selection was based more on plausibility and analogy than solid evidence. It was fairly clear that variations like those observed in domestic animals brought about some changes in nature; Darwin extrapolated to assert that all differences between living creatures were thus caused, ultimately back to the separation of humans, fish, protozoa, and plants. In order to exclude anything savoring of divine intervention, Darwin also assumed that change had to be gradual and random.

The Genetic Factor

A serious weakness of the theory was ignorance regarding how variations are transmitted. Darwin theorized that particles from all parts of the body gather in the reproductive organs to determine the inheritance of the offspring (pangenesis), but he had no evidence. The theory also had the defect that it led back to Lamarck's earlier idea, which Darwin wanted to supersede, that evolution progresses by the inheritance of acquired traits. In the conventional example, the giraffe came about because a gazellelike animal stretched to reach higher foliage, its offspring were born with longer necks, and, after many generations, voilà: a giraffe.

Without knowledge of genetics, it was difficult to understand how random variations could change a species. Any rare improvement would be diluted by half through mating with an animal without the improvement, and it would be lost in the general population, critics argued, in a few generations. . . .

There seemed to be no answer until the work of Gregor Mendel. Darwin, however, ignored this work when he published his findings in 1866. . . . But in 1900 three investigators rediscovered Mendel's theory that traits are inherited in simple unitary fashion; hence they are not diluted by mating and reproduction but can reappear in subsequent generations. This meant that the features making up the plant or animal are mechanically transmitted and are more or less independent of the rest of the organism. It became difficult to conceive of acquired characteristics being somehow rather mysteriously incorporated into the genetic makeup, or genotype, of the individual and species. . . .

Mendelian genetics initially seemed to contradict Darwin's gradualist variation. But in the first decades of [the twentieth] century, biologists melded Mendelism with Darwinism to make what came to be regarded as a definitive synthesis of evolutionary theory. In a research binge, geneticists working with cages of fruit flies discovered [enough complex evidence] . . . to construct what seemed to be a solid theory of evolution. The synthesis completed in the 1940s became an imposing theoretical edifice.

This success, however, was more imposing than solid. Genetics proved to be unexpectedly and confusingly complicated. Mendel's experiments had worked out well partly because he had extraordinarily good luck: the several genes that he traced through generations of peas happened to occur on different chromosomes and manifested themselves in simpler fashion than usually occurs. Genes were found to combine variously, to have multiple effects, and to change, or mutate, in ways difficult to explain. Moreover, the mountain of data produced little understanding of evolution. Fruit flies had mutations of eye color or even sprouted a leg from the head, but nothing was learned about how new organs originated and little about how different species . . . may have arisen.

Chance Produces Novelty

Consequently, many biologists looked for forces or directions in evolution beyond those postulated by the basic Darwinist theory. Some resorted to vitalism—the theory that living beings are alive by virtue of a special life force—or to some variant of Lamarckism, or to a teleology—the belief in a built-in purposiveness of evolution.

The variation-selection theory was refortified, however, by another big advance of genetics. The discovery in the 1950s that nucleic acid (DNA and RNA) is the carrier of heredity, demonstrating the physical materiality of the genes, restored confidence in the materialistic explanation of evolution. Because nucleic acid both replicated itself and transmitted information to proteins, molecular biologists concluded that information flows only from the nucleic acid to the body, never in the reverse direction. This implied that changes in the genetic apparatus, or genome, could come about only by errors of replication. Variation had to be accidental; there could be feedback only through the relative success or failure of variants in propagating themselves. This beautifully mechanical-materialistic idea was hardly to be questioned. As a biochemist put it, "Advances in understanding of the details of genetics [through discovery of DNA] confirmed Darwin's theory and documented the processes to the same extent that Newton's laws of physics had been validated."

The successes of the new molecular biology in analyzing and manipulating enzymes and nucleic acids made it easy to believe that unraveling the material substance of heredity would reveal the secrets of evolution. Again, however, answers remained elusive. Relations between genes and organism turned out to be excessively complicated, and the voluminous information coming out of the laboratories was of little help in interpreting the fossil record or understanding structures and behaviors. But the treatment of heredity in concrete molecular terms set the tone of the discussion for most practicing biologists. . . .

An unconditional thesis of neo-Darwinism is that "Chance is the only source of true novelty." This means that innovations spring from errors in the reproductive process. A large majority of the mistakes are insignificant or harmful, dislocations in a well-organized and tested apparatus, but these are weeded out by the stern culling of the environment. A few errors are improvements, enabling the organism to cope better and reproduce itself more abundantly, especially when environmental changes raise new demands. As variations accumulate, the interbreeding population changes indefinitely, eventually altering the species or making a new species. Darwin assumed that variation could tend in any direction without definite limits, and the modern theory tends to favor this assumption, which leaves the organism subject to the roulette wheel of variation and the pressures of its world. . . .

The important point is that there can be nothing purposive or teleological in evolution; any notion of inherent purpose would make nature less amenable to objective analysis. For a biologist to call another a teleologist is an insult. Even the idea of direction in evolution caused by internal factors, or orthogenesis, is disliked. The sole force for change must be adaptation.

Many biologists go on to refuse to recognize any overall direction in evolution. They even dislike the notion that some creatures are in any important way "higher" than others. In spite of the fact that natural selection implies improvement and that a mammal is much further from its presumed one-celled ancestor than is an amoeba, they sense a contradiction between the idea of "higher" forms and mechanistic means of change. . . .

The "Selfish Gene"

Conventional theorists also prefer to think as much as possible in terms of the material particles of heredity, the genes. For Darwin, the unit of selection was the individual, which might be enabled by its qualities (plus good luck) to give rise to more than its share of the new generation. The discoveries of genetics led to a shift of emphasis, however, from the organism to its reproductive material. In the view of population genetics, as developed by Sewall Wright, Ronald Fisher, and others, a species is characterized simply by the frequency of various genes and their alternatives at any given site. . . . Evolutionary change thus was reduced to a matter of gene statistics. . . . The product is a set of impressive and sometimes useful equations governing the probable alteration of a population (that is, evolutionary change), equations that are sometimes taken as the biological equivalent of Newton's laws.

Population genetics is less firm, however, than classical mechanics. Its chief variable, gene frequency, is seldom measurable in practice; its principal independent variable, fitness, can only be guessed because it is impossible to determine to what degree survival is a matter of special genes or accident or special circumstances. More broadly, an organism cannot be treated simply as the product of a number of proteins, each produced by the corresponding gene. Genes have multiple effects, and most traits depend on multiple genes. The selection of individual genes is most important in very simple organisms. That is, population genetics is best applicable to bacteria, and it does not tell much about the evolution of organs and higher animals. . . .

The stark affirmation of the "selfish gene" appeals for its counterintuitive boldness. But to say that the genes are in some indefinable way primary is more of an ideological than a scientific statement. Genes are not independent entities but dependent parts of an entirety that gives them effect. All parts of the cell interact, and the combinations of genes are at least as important as their individual effects in the making of the organism. Selection operates not on genes but on organisms or perhaps groups (and possibly species)....

The selection-variation theory is a useful approximation. It doubtless accounts for, or helps to account for, very much. Natural selection serves to eliminate those less qualified in the competition; that is, it stabilizes forms. Darwin, however, thought of selection not so much as elimination of the unfit as the opportunity of the fitter, that is, adaptive change.

The best studied and most frequently cited example is the alteration of the British peppered moth (*Biston betularia*). In this species, as in many other species of moths and other insects, a dark-colored variety has taken over in areas where industrial smoke has blackened much of the landscape, replacing a mottled gray form that was camouflaged on lichen-covered surfaces. This adaptation is selected for by natural agents; birds have difficulty spotting dark-colored moths on a dark background. A clean air act was followed by a return of lighter-colored varieties. But things are not simple. The dark (carbonaria) gene was not fully dominant at first but became so. Melanic moths have invaded rural areas with lichen-covered tree trunks; it seems that they may have become more viable for reasons other than camouflage. The victory of the darker variety is interesting, but it does not prove that such a selective process can account for the ability of chameleons to camouflage themselves by reflex in a few minutes.

Variants of the Theory

To meet problems abundantly arising, many biologists, while adhering in principle to Darwinism, have suggested modifications to the basic theory, and evolutionary doctrine has become less coherent. In principle it is held correct, but when one tries to apply it—as in the question of human origins—controversy arises.

A mild challenge to the theory is the idea that populations change not only through adaptation but through mutations that are neutral or at least not seriously negative. Strict adaptationists claim that there "must be an evolutionary advantage to a trait if we only look hard enough." But for each positive change, there must be thousands or tens of thousands that are not clearly useful, and unuseful traits (except for those harmful enough to be eliminated by selection) can lead to changes of the population as variants are increased or eliminated by the workings of pure chance. Suppose one has a barrel filled with a mixture of black beans and white beans. After stirring it, one removes half and throws them away, and then doubles the other half and returns it to the barrel, like a new generation being born, and repeats the operation. The proportions of black and white will change in each "generation." After enough times, one will have all black or all white beans. How many "generations" it will take depends on luck and the number of beans.

Such change, called "genetic drift" by Sewall Wright, takes part of the honor of driving evolution away from the principle of adaptation. That drift is a reality is indicated by the universality of molecular change in proteins and its greater frequency where selection is less important. . . .

A sharper challenge to mainstream theory is the rejection of gradualism in favor of "stasis-punctuation," the idea that change does not simply flow along but is sharply punctuated. Darwin insisted (despite an occasional contrary remark) on gradualism, although his theory did not really require it. The continuity of nature was a major discovery of his age; the corollary of present-day continuity was that the past was to be understood in terms of forces at work in the present. Attributing change of species to an accumulation of small variations made everything seem most natural and understandable. This idea also fitted Darwin's philosophical outlook: gradualism was the opposite of the special creationism that he combated.

The conviction that change must have been gradual harmonizes with the rejection of any idea of purpose in evolution. If organisms grope their way, so to speak, into new adaptations, they do it by little steps. Hence, biologists for the most part prefer the gradualism that Darwin preferred. But the fossil record fails to show continuous series linking different groups, and some biologists early in this century, led by Richard Goldschmidt, attempted to show that evolution must have proceeded by radical leaps—in Goldschmidt's phrase, "hopeful monsters." This idea was pretty much abandoned. It was argued that no such leaps have been observed, and when gross mutational changes do appear in the laboratory, such as a doubled thorax in a fruit fly or a leg coming out of the head, they consist of errors in the formation or placement of old parts, never the appearance of a new organ. . . .

Recently, however, some theorists, led by Stephen J. Gould and Niles Eldredge, have forcefully argued on a more sophisticated level that evolution is by no means a mere accumulation of tiny, continually occurring changes. They contend that there is a dichotomy [division] between stability of species and rapid change, or "stasis and punctuation." More conventional theorists dislike this non-Darwinian thesis because it seems to contradict the strictly mechanistic approach, at least in spirit. . . .

Traits Good for the Group

Evolutionists also differ in their emphasis on single genes or combinations, on genes that make proteins or on genes that turn other genes on or off. If most traits are the effect of multiple genes, regulatory genes, or gene-enzyme systems, large changes become more conceivable as results of recombinations, and mutations to make new proteins are correspondingly less important. The process of inheritance is evidently much more complicated than appeared from Mendel's tracing simple, unitary traits in peas—

smooth or wrinkled, yellow or green, dwarf or tall. This implies shifting emphasis from relatively well-understood structural genes (which produce not structures but proteins that may or may not go into structures) to the poorly understood regulatory genes, which turn batteries of genes on and off. No simple theory can cope with the enormous complexity revealed by modern genetics.

A different problem is the existence of many traits that are evidently advantageous for the group but not for the individual. One prairie dog stands guard while others feed, forgoing a meal and exposing itself to predators. A honeybee not only rushes out to defend the hive but sacrifices its life by leaving its barbed stinger in the flesh of the intruder, along with the poison sac and part of the bee's guts. To resolve this contradiction, evolutionists have postulated that selection operates not only on the basis of individual but of group advantage: a group possessing traits useful for its collective success would prosper and reproduce, eventually perhaps replacing groups lacking the socially useful trait.

On the other hand, individuals lacking the altruistic gene would presumably have greater individual reproductive success, and the social trait would be wiped out. To overcome this contradiction, William Hamilton in the 1960s and 1970s elaborated the idea of indirect fitness, a significant addition to the Darwinist canon. An animal has "direct fitness" in its ability to propagate its lineage; it also has "indirect fitness" insofar as it helps other individuals with which it is related by sharing genes. This idea has proved handy in evolutionary theory despite its logical weaknesses.

Evolutionary theory may be modified to meet such difficulties, and evolutionists differ widely in their views regarding the pace, focus, and mechanics of change. They firmly maintain, however, the central ideas: there is nothing purposive, and organisms adapt genetically only by success or failure in leaving descendants. In the words of Ernst Mayr, "The one thing about which modern authors are unanimous is that adaptation is not teleological."

The Commitment to Darwinism

Darwin answered the intellectual need of the day, and the age recognized itself in him. He has been elevated as perhaps the greatest of scientists, and his name stands for a theory that has grown far beyond his work. What is commonly called the neo-Darwinian synthesis, or simply the modern synthesis, has taken on somewhat ideological overtones, especially in the United States. It becomes a little like a revelation by a prophet, whose every word in his major works is recorded in concordances. . . .

In a common view, the accepted evolutionary doctrine, rough hewn as it may be, has to be regarded as true unless it is proved false, even though the evidence for it is admittedly incomplete. Mark Ridley, for example, again and again makes the case for natural selection simply on the grounds that we have no other plausible explanation. This perspective is understandable, perhaps persuasive. Theories in which many scientists have invested their careers are not set aside until they can be replaced by more satisfactory theories, usually brought forward by younger thinkers. Neo-Darwinism is an accepted "mode of cognition" as conceived by historians of science.

Science advances by testing, modifying, and so far as necessary replacing hypotheses, but standard evolutionary theory is not usually treated as a hypothesis to be investigated. A single counterexample refutes a mathematical theorem, but evolutionary theory is in practice not falsifiable. Many very simple facts, such as that all the millions of species of insects and no species of noninsects have six legs, might well be considered to disprove natural selection as a generalization. But such broad problems are usually ignored, and it is assumed that any puzzle must be solvable in its terms if adequately studied. . . .

Despite the infrequency of any useful mutation, it can always be postulated that the appropriate mutations came along by accident and were selected, bringing about the adaptation in question. For example, it is hypothesized that

natural selection has led the female sedge warbler to prefer full-throated males because they should make good for-
agers for the family. On the other hand, the female lyrebird
supposedly has been selected to prefer the male who ne-
glects his offspring and so avoids bringing the nest to the
attention of predators. The female spotted hyena, in the
opinion of some, has a set of external genitals like those of
the male in order the better to greet her friends. Some
weaverbirds are monogamous because food is scarce, others because food is abundant. Marmot families stay together longer at high altitudes because there is less vegetation. If the young ones dispersed sooner at high altitudes,
it would probably be because where food is scarce they
have to seek new pastures. . . .

In practice, however, biologists are usually rather real-
istic. They think of more evolved plants and animals as "advanced" or "higher," and they use those words. However
scrupulously they avoid teleological language, they recognize certain directions in evolution. They are likely to speak
of animals having purposes and acting to secure ends, if
only because it is awkward to say that the animal has
genes causing its limbic system to direct certain actions
conducive to its reproductive success. Having devotedly
spent thousands of hours observing lions or jays or dam-
selflies doing nothing in particular, they faithfully report
facts that do not accord with the standard theory.

It is comforting, however, to see nature as basically me-
chanical, and hence totally understandable. "To maximize
fitness" is an offhand explanation for almost anything, usually persuasive until critically examined. . . .

The principle of variation-selection represents a mental
economy and suggests a way to seek answers, a key to unraveling the infinite variety and complexity of living nature.
It gives a clear-cut orientation, whether or not its explanations are adequate.

No other science has such a comfortable foundation. The
physicists had something like it in the Newtonian synthe-

sis, but they lost it because they learned too much. Sociologists and other social scientists (except for devotees of certain schools) have never been able to cherish even an illusion of total intellectual mastery. But the neo-Darwinian synthesis gives biologists a satisfying basis on which to work, most suitable for those who simply want to get on with their interesting investigations. The variation-selection theory even offers an uncomplicated view of the human condition and promises a means of coming to grips with the bafflements of human nature. The doctrine being axiomatically true, there must be a basis in natural selection for all human as well as other animal behavior; we have only to search for it. Sociobiologists find satisfaction in the bold assertion of a firmly scientific attitude in a controversial area where no one else can offer satisfactory answers. . . .

There are strong reasons for reluctance to admit any modification of evolutionary theory that might lead away from its mechanistic essence. Humans have adhered much more blindly to many less rational beliefs.

Punctuation and Direct Mutation: Challenges to Darwinism?

GARY CZIKO

Among the many new developments in evolutionary theory in the second half of the twentieth century were the concepts of punctuated equilibrium and directed mutation. The first, proposed by scientists Stephen Jay Gould and Niles Eldredge in the 1970s, suggests that evolution often works in sudden, relatively rapid bursts. The second, suggested by a team led by molecular biologist John Cairns in the 1980s, makes the case that organisms might partially direct their own evolution rather than rely solely on natural selection. This useful overview of the two theories and how they have challenged the field of evolutionary biology is by University of Illinois scholar Gary Cziko.

Universal selection theory draws heavily on biological evolution for its inspiration. Although the evolution of living forms is only one instance of a selectionist process resulting in adapted complexity, it provides the foundation and inspiration for all other selectionist theories of the emergence of fit. Biology has also lived longer with selectionist thinking than any other discipline. For these reasons, recent developments in biology that cast doubts on the fundamental role of natural selection in the emergence of the adapted complexity of living organisms are of considerable interest to those attempting to extend the selectionist perspective to other fields. If cumulative blind variation and selection is found to be lacking as an explanation for the emergence of design in organic evolution, an extension of selectionist principles to other achievements of adapted complexity would be suspect. We will therefore now confront [some] would-be challengers to natural selection: punctuated equilibrium [and] directed mutation. . . .

Punctuated Equilibrium

According to classic Darwinian selection, biological evolution proceeds through the accumulation of very small changes over long periods of time. Gradual change is essential, since it is the only way that blind variation is likely to come up with improvements for selection. Whereas it is always possible that a large genetic change (or macromutation) may result in a fitter organism, for example, the transformation of an organism completely insensitive to light to one with a functioning eye in one generation, the laws of probability are almost certain to make large random changes less adaptive rather than more. As [noted scientist Richard] Dawkins explains:

> To "tame" chance means to break down the very improbable into less improbable small components arranged in series. No matter how improbable it is that an X could have arisen from Y in a single step, it is always possible to conceive of a series of infinitesimal graded intermediates between them. However improbable a large-scale change may be, smaller changes are less improbable. And provided we postulate a sufficiently large series of sufficiently finely graded intermediates, we shall be able to derive anything from anything else, without astronomical improbabilities.

The problem is, however, that the fossil record does not

provide clear evidence for the gradual change of even one species into another. Darwin recognized the incompleteness of the fossil record, but believed that it was only a matter of time before these intermediate "missing links" would be found to provide hard evidence for the gradual emergence of new species over time. That these fossil gaps remain despite many new fossil finds has been taken by some as an indication that Darwin's emphasis on the gradualism of evolution was mistaken, and that evolution proceeds not by slow, gradual changes but rather by large and dramatic jumps, or saltations.

True saltationists are not easy to find among modern evolutionary biologists, since it is generally recognized that large, blind macromutations from parent to offspring are almost certain to be maladaptive. But a well-known antigradualist perspective is present today in the theory of "punctuated equilibrium," developed by Gould and Eldredge.

These researchers theorize that instead of continuous gradual change over time, the evolution of a species is marked by long periods of no or little change (stasis) interrupted occasionally by short periods of relatively rapid evolutionary change (punctuations). This may be a somewhat different picture of evolution than originally conceived by Darwin, but it is not inconsistent with the gradualism that is an essential part of natural selection. Although punctuated equilibrium describes relatively rapid change, this change still takes place over very long time periods, in the range of many thousands of years to much longer.

What is characteristic of punctuated equilibrium, then, is not the belief in adaptive macromutations arising in a single generation, but rather the long periods of stasis. But these periods need not be considered mysterious since they may simply be an indication that the species was already well adapted to its environment, and that the environment was not undergoing any rapid changes that would have created new selection pressures requiring new adaptations. So actually nothing in the theory of punctuated equilibrium is in any way fundamentally inconsistent with Darwin's conception of evolution.

Directed Mutation

Similar compatibility is not the case, however, for another view of evolution, that has attracted considerable interest and led to much recent controversy. In 1988 John Cairns, a well-respected molecular biologist and cancer researcher, published with two associates a paper in the prestigious British journal *Nature* that threatened to undermine the basic tenets of Darwinian evolution.

Cairns and his colleagues claimed to have found evidence that [the bacterium] *E. coli* was able somehow to direct its mutations to achieve adaptive changes when placed in a new, challenging environment. This research involved placing bacteria that could only metabolize glucose in an environment where only a foreign sugar (lactose) was available. Here the stressed bacteria continued to duplicate and, as would be expected, some of the descendants contained mutations that permitted them to metabolize the new sugar. This in itself is not surprising, since the genetic change necessary to transform an *E. coli* from a glucose- to a lactose-eating bacterium is quite small, and in a large colony it would be expected that at least some of the naturally occurring mutants would have stumbled on it by sheer blind chance. But these scientists reached the highly unorthodox conclusion that instead of being produced randomly, the bacteria were somehow able to produce the adaptive mutations at a much higher frequency than other, nonadaptive mutations. In other words, they believed that their studies provided evidence that "bacteria can choose which mutations they should produce" which would "provide a mechanism for the inheritance of acquired characteristics."

As would be expected, these statements immediately elicited both considerable interest and controversy, since the central dogma of biology was being challenged, that is, that changes in the environment cannot direct (instruct) changes in the genome. Some researchers rejected this conclusion out of hand, but others were impressed enough to attempt to find possible mechanisms by which the environment could somehow instruct the genome to produce just the right mutations to allow it to digest the new sugar. Cairns himself proposed that environmental changes could affect changes in proteins that could consequently instruct the DNA to make certain adaptive changes in the genes, in flagrant violation of the central dogma.

However, it may well be that this and other explanations for directed or "instructed" mutation are not necessary after all. Australian microbiologist Donald MacPhee and his colleagues provided evidence that, when placed in a medium of lactose, the mutations produced by glucose-metabolizing *E. coli* are indeed produced blindly. What seems to happen under the stressed condition of a glucose-poor environment is not a specific increase in the rate of adaptive mutations, but rather a general increase in the overall mutation rate due to inhibition of the mechanism that usually checks and repairs the genetic errors that arise during the normal functioning of the bacterium. So while mutations continue to be produced blindly, the higher rate of genetic change allows the bacteria to stumble on the adaptive genetic change more quickly than they would if left in their normal glucose-rich environment.

But let us continue to imagine for a moment that a bacterium was able to change just those genes regulating metabolism in just the right way to allow for the digestion of a foreign sugar. If this were the case, it would be yet another example of a puzzle of fit demonstrating that the bacterium had somehow acquired the ability to sense a new sugar in its environment and alter its genome to digest it. But then we would be led to ponder how this adapted complexity could have originated in the first place, with cumulative blind variation and selection as a prime candidate to explain the source of this remarkable ability that somehow permitted the bacterium to instruct its genome to make the required changes to digest the new, strange food that was being served.

Organisms in Control of Evolution?

Although no convincing evidence exists that adaptive changes in genes can be directed by the environment . . . the findings of Cairns and MacPhee and their respective colleagues are important. If organisms are able to increase their mutation rate in the presence of new environmental stresses but keep mutations in check when these stresses are absent, it would enable organisms to exert a certain degree of control over evolution that is absent from the classic neo-Darwinian perspective. Instead of producing mutations at a constant rate regardless of environmental conditions, organisms may produce more mutations and therefore more varied offspring just when such innovative variation is necessary to keep the species extant.

This view ascribes to the evolutionary process decidedly more "intelligence" than does the neo-Darwinian perspective. It nonetheless preserves the required blindness of genetic variations. What is altered is only the rate of production of these variations. This sensitivity of mutation rate to environmental stress could simply be the result of a stress-related breakdown of genetic repair mechanisms. Or it could be the result of a more sophisticated active mechanism that itself had evolved by natural selection, since individuals that by chance produced more genetic variability under difficult environmental conditions would have been more likely to leave better adapted progeny than those insensitive to environmental stress.

The work of Cairns and MacPhee concerned the metabolism of different types of food. It is not difficult to imagine how other types of biological functions could also be involved, such as thermoregulation. For example, as temperatures dropped at the onset of an ice age, mammals would undergo stress as did Cairns's bacteria when placed in an environment where no useful food was available. This would lead to an increase in the mutation rate during reproduction, resulting in a second generation of animals with greater variation in the length and texture of their coats. Those particular descendants having, by chance, longer and thus warmer coats would suffer less from the cold environment, resulting in lower mutation rates and consequently less variation in the coats of their third generation, extra-hairy offspring. But those second-generation animals with short coats would maintain a higher rate of mutation, so that at least some of their offspring would likely have warmer coats than their parents.

This hypothesis has some interesting consequences. As in the ice-age example, by varying the mutation rate, a species would adapt more quickly to changing environmental conditions. It is also of interest to realize that such stress-dependent mutation rates would result in occasional short periods of relatively rapid (although still gradual) evolutionary change separated by longer periods of little or no change during periods of environmental stability. And this is exactly what Gould, Eldredge, and their associates refer to as punctuated equilibrium.

Darwinian Evolution May Not Work on the Molecular Level

MICHAEL J. BEHE

One of the more recent books about evolutionary theory is Darwin's Black Box, *by Michael J. Behe, a professor of biochemistry at Lehigh University. In this excerpt, Behe makes the main point of his thesis, namely that, although natural selection seems to work on the larger level of animal organs and populations, it is difficult to detect, and perhaps does not work, on the microscopic level of molecules and atoms. The question for Behe is: Where did the complex chemicals that make up animal organs—for instance, the light receptors in the eyes—come from? This, he points out, may be one of the biggest challenges faced by modern evolutionary theory.*

L ike many great ideas, Darwin's is elegantly simple. He observed that there is variation in all species: some members are bigger, some smaller, some faster, some lighter in color, and so forth. He reasoned that since limited food supplies could not support all organisms that are born, the ones whose chance variation gave them an advantage in the struggle for life would tend to survive and reproduce, outcompeting the less favored ones. If the variation were inherited, then the characteristics of the species would change over time; over great periods, great changes might occur.

For more than a century most scientists have thought that virtually all of life, or at least all of its most interesting features, resulted from natural selection working on random variation. Darwin's idea has been used to explain finch beaks and horse hoofs, moth coloration and insect slaves, and the distribution of life around the globe and through the ages. The theory has even been stretched by some scientists to interpret human behavior: why desperate people commit suicide, why teenagers have babies out of wedlock, why some groups do better on intelligence tests than other groups, and why religious missionaries forgo marriage and children. There is nothing—no organ or idea, no sense or thought—that has not been the subject of evolutionary rumination.

Almost a century and a half after Darwin proposed his theory, evolutionary biology has had much success in accounting for patterns of life we see around us. To many, its triumph seems complete. But the real work of life does not happen at the level of the whole animal or organ; the most important parts of living things are too small to be seen. Life is lived in the details, and it is molecules that handle life's details. Darwin's idea might explain horse hoofs, but can it explain life's foundation?

The Case of Molecular Machines

Shortly after 1950 science advanced to the point where it could determine the shapes and properties of a few of the molecules that make up living organisms. Slowly, painstakingly, the structures of more and more biological molecules were elucidated, and the way they work inferred from countless experiments. The cumulative results show with piercing clarity that life is based on *machines*—machines made of molecules! Molecular machines haul cargo from one place in the cell to another along "highways" made of other molecules, while still others act as cables, ropes, and pulleys to hold the cell in shape. Machines turn cellular switches on and off, sometimes killing the cell or causing it to grow. Solar-powered machines capture the energy of photons and store it in chemicals. Electrical machines allow current to flow through nerves. Manufacturing machines build other molecular machines, as well as themselves. Cells swim using machines, copy themselves with machinery, ingest food with machinery. In short, highly sophisticated molecular machines control every cellular process. Thus the details of life are finely calibrated, and the machinery of life enormously complex.

Can all of life be fit into Darwin's theory of evolution? Because the popular media likes to publish exciting stories, and because some scientists enjoy speculating about how far their discoveries might go, it has been difficult for the public to separate fact from conjecture. To find the real evidence you have to dig into the journals and books published by the scientific community itself. The scientific literature reports experiments firsthand, and the reports are generally free of the flights of fancy that make their way into the spinoffs that follow. But as I will note later, if you search the scientific literature on evolution, and if you focus your search on the question of how molecular machines—the basis of life—developed, you find an eerie and complete silence. The complexity of life's foundation has paralyzed science's attempt to account for it; molecular machines raise an as-yet-impenetrable barrier to Darwinism's universal reach. . . .

With the advent of modern biochemistry we are now able to look at the rock-bottom level of life. We can now make an informed evaluation of whether the putative small steps required to produce large evolutionary changes can ever get small enough. . . . The canyons separating everyday life forms have their counterparts in the canyons that separate biological systems on a microscopic scale. Like a fractal pattern in mathematics, where a motif is repeated even as you look at smaller and smaller scales, unbridgeable chasms occur even at the tiniest level of life.

Biochemistry has pushed Darwin's theory to the limit. It

has done so by opening the ultimate black box, the cell, thereby making possible our understanding of how life works. It is the astonishing complexity of subcellular organic structures that has forced the question, How could all this have evolved? To feel the brunt of the question— and to get a taste of what's in store for us—let's look at an example of a biochemical system. An explanation for the origin of a function must keep pace with contemporary science. Let's see how science's explanation for one function, vision, has progressed since the nineteenth century, then ask how that affects our task of explaining its origin.

In the nineteenth century, the anatomy of the eye was known in detail. The pupil of the eye, scientists knew, acts as a shutter to let in enough light to see in either brilliant sunlight or nighttime darkness. The lens of the eye gathers light and focuses it on the retina to form a sharp image. The muscles of the eye allow it to move quickly. Different colors of light, with different wavelengths, would cause a blurred image, except that the lens of the eye changes density over its surface to correct for chromatic aberration. These sophisticated methods astounded everyone who was familiar with them. Scientists of the nineteenth century knew that if a person lacked any of the eye's many integrated features, the result would be a severe loss of vision or outright blindness. They concluded that the eye could function only if it were nearly intact.

Charles Darwin knew about the eye, too. In *The Origin of Species* Darwin dealt with many objections to his theory of evolution by natural selection. He discussed the problem of the eye in a section of the book appropriately entitled "Organs of Extreme Perfection and Complication." In Darwin's thinking, evolution could not build a complex organ in one step or a few steps; radical innovations such as the eye would require generations of organisms to slowly accumulate beneficial changes in a gradual process. He realized that if in one generation an organ as complex as the eye suddenly appeared, it would be tantamount to a miracle. Unfortunately, gradual development of the human eye appeared to be impossible, since its many sophisticated features seemed to be interdependent. Somehow, for evolution to be believable, Darwin had to convince the public that complex organs could be formed in a step-by-step process.

He succeeded brilliantly. Cleverly, Darwin didn't try to discover a real pathway that evolution might have used to make the eye. Rather, he pointed to modern animals with different kinds of eyes (ranging from the simple to the complex) and suggested that the evolution of the human eye might have involved similar organs as intermediates.

Here is a paraphrase of Darwin's argument: Although humans have complex camera-type eyes, many animals get by with less. Some tiny creatures have just a simple group of pigmented cells—not much more than a light-sensitive spot. That simple arrangement can hardly be said to confer vision, but it can sense light and dark, and so it meets the creature's needs. The light-sensing organ of some starfishes is somewhat more sophisticated. Their eye is located in a depressed region. Since the curvature of the depression blocks off light from some directions, the animal can sense which direction the light is coming from. The directional sense of the eye improves if the curvature becomes more pronounced, but more curvature also lessens the amount of light that enters the eye, decreasing its sensitivity. The sensitivity can be increased by placement of gelatinous material in the cavity to act as a lens; some modern animals have eyes with such crude lenses. Gradual improvements in the lens could then provide increasingly sharp images to meet the requirements of the animal's environment.

Using reasoning like this, Darwin convinced many of his readers that an evolutionary pathway leads from the simplest light-sensitive spot to the sophisticated camera-eye of man. But the question of how vision began remained unanswered. Darwin persuaded much of the world that a modern eye evolved gradually from a simpler structure, but he did not even try to explain where his starting point—the relatively simple light-sensitive spot—came from. On the contrary, Darwin dismissed the question of the eye's ultimate origin: "How a nerve comes to be sensitive to light hardly concerns us more than how life itself originated."

He had an excellent reason for declining the question: it was completely beyond nineteenth-century science. How the eye works—that is, what happens when a photon of light first hits the retina—simply could not be answered at that time. As a matter of fact, no question about the underlying mechanisms of life could be answered. How did animal muscles cause movement? How did photosynthesis work? How was energy extracted from food? How did the body fight infection? No one knew.

The Vision of Biochemistry

To Darwin, vision was a black box, but after the cumulative hard work of many biochemists, we are now approaching answers to the question of sight. The following five paragraphs give a biochemical sketch of the eye's operation. Don't be put off by the strange names of the components. They're just labels, no more esoteric than *carburetor* or *differential* are to someone reading a car manual for the first time. . . .

When light first strikes the retina a photon interacts with a molecule called 11-*cis*-retinal, which rearranges within picoseconds to *trans*-retinal. (A picosecond is about the time it takes light to travel the breadth of a single human hair.) The change in the shape of the retinal molecule forces a change in the shape of the protein, rhodopsin, to which the retinal is tightly bound. The protein's metamorphosis alters its behavior. Now called metarhodopsin II, the protein sticks to another protein, called transducin. Before bumping into metarhodopsin II, transducin had tightly bound a small molecule called GDP. But when transducin interacts with metarhodopsin II, the GDP falls off, and a molecule called GTP binds to transducin. (GTP is closely related to, but critically different from, GDP.)

GTP-transducin-metarhodopsin II now binds to a protein called phosphodiesterase, located in the inner membrane of the cell. When attached to metarhodopsin II and its entourage, the phosphodiesterase acquires the chemical ability to "cut" a molecule called cGMP (a chemical relative of both GDP and GTP). Initially there are a lot of cGMP molecules in the cell, but the phosphodiesterase lowers its concentration, just as a pulled plug lowers the water level in a bathtub.

Another membrane protein that binds cGMP is called an ion channel. It acts as a gateway that regulates the number of sodium ions in the cell. Normally the ion channel allows sodium ions to flow into the cell, while a separate protein actively pumps them out again. The dual action of the ion channel and pump keeps the level of sodium ions in the cell within a narrow range. When the amount of cGMP is reduced because of cleavage by the phosphodiesterase, the ion channel closes, causing the cellular concentration of positively charged sodium ions to be reduced. This causes an imbalance of charge across the cell membrane that, finally, causes a current to be transmitted down the optic nerve to the brain. The result, when interpreted by the brain, is vision.

If the reactions mentioned above were the only ones that operated in the cell, the supply of 11-*cis*-retinal, cGMP, and sodium ions would quickly be depleted. Something has to turn off the proteins that were turned on and restore the cell to its original state. Several mechanisms do this. First, in the dark the ion channel (in addition to sodium ions) also lets calcium ions into the cell. The calcium is pumped back out by a different protein so that a constant calcium concentration is maintained. When cGMP levels fall, shutting down the ion channel, calcium ion concentration decreases, too. The phosphodiesterase enzyme, which destroys cGMP, slows down at lower calcium concentration. Second, a protein called guanylate cyclase begins to resynthesize cGMP when calcium levels start to fall. Third, while all of this is going on, metarhodopsin II is chemically modified by an enzyme called rhodopsin kinase. The modified rhodopsin then binds to a protein known as arrestin, which prevents the rhodopsin from activating more transducin. So the cell contains mechanisms to limit the amplified signal started by a single photon.

Trans-retinal eventually falls off of rhodopsin and must be reconverted to 11-*cis*-retinal and again bound by rhodopsin to get back to the starting point for another visual cycle. To accomplish this, *trans*-retinal is first chemically modified by an enzyme to *trans*-retinol—a form containing two more hydrogen atoms. A second enzyme then converts the molecule to 11-*cis*-retinol. Finally, a third enzyme removes the previously added hydrogen atoms to form 11-*cis*-retinal, a cycle is complete.

The Lilliputian Challenge to Darwin

The above explanation is just a sketchy overview of the biochemistry of vision. Ultimately, though, *this* is the level of explanation for which biological science must aim. In order to truly understand a function, one must understand in detail every relevant step in the process. The relevant steps in biological processes occur ultimately at the molecular level, so a satisfactory explanation of a biological phenomenon—such as sight, digestion, or immunity—must include its molecular explanation.

Now that the black box of vision has been opened, it is no longer enough for an evolutionary explanation of that power to consider only the *anatomical* structures of whole eyes, as Darwin did in the nineteenth century (and as popularizers of evolution continue to do today). Each of the anatomical steps and structures that Darwin thought were so simple actually involves staggeringly complicated biochemical processes that cannot be papered over with rhetoric. Darwin's metaphorical hops from butte to butte are now revealed in many cases to be huge leaps between carefully tailored machines—distances that would require a helicopter to cross in one trip.

Thus biochemistry offers a Lilliputian challenge [a reference to the Lilliputians, tiny people in Jonathan Swift's *Gulliver's Travels*] to Darwin. Anatomy is, quite simply, irrelevant to the question of whether evolution could take place on the molecular level. So is the fossil record. It no longer matters whether there are huge gaps in the fossil record or whether the record is as continuous as that of U.S. presidents. And if there are gaps, it does not matter whether they can be explained plausibly. The fossil record has nothing to tell us about whether the interactions of 11-*cis*-retinal with rhodopsin, transducin, and phosphodiesterase could have developed step-by-step. Neither do the patterns of biogeography matter, nor those of population biology, nor the traditional explanations of evolutionary theory for rudimentary organs or species abundance. This is not to say that random mutation is a myth, or that Darwinism fails to explain anything (it explains microevolution very nicely), or that large-scale phenomena like population genetics don't matter. They do. Until recently, however, evolutionary biologists could be unconcerned with the molecular details of life because so little was known about them. Now the black box of the cell has been opened, and the infinitesimal world that stands revealed must be explained.

DNA Offers New Clues to Human Evolution

STEVE JONES

Charles Darwin and his contemporaries knew nothing about DNA and the role it plays in heredity, of course. And the discovery and study of DNA in the twentieth century revised the way scientists viewed the process of evolution. In this essay, Steve Jones, a professor of genetics at University College London, first makes the point that it is the complexity of the human brain that separates people from lower animals; and it is in the brain that molecules of DNA are most diverse and complex. He then explains how scientists can now use traces of DNA in past human remains as markers to trace the evolution of the human species, as well as to follow ancient human population movements across the globe.

Human behavior is, without doubt, crafted by evolution, and we have inborn instincts as strong as those of any animal. Our species has long been seen as an ape remade by thought: to one expert, "The ancestors of the Gorilla and the Chimpanzee gave up the struggle for mental supremacy because they were satisfied with their circumstances." Humans, it appears, were not. The brain makes us human. It has doubled in size in the past two million years. Its progress was not smooth, as some of our forerunners had brains larger than our own (albeit on a heftier

Steve Jones, *Darwin's Ghost: The Origin of Species Updated.* New York: Random House, 1999. Copyright © 1999 by Steve Jones. Reproduced by permission of the author and the author's agents, Scovil Chichak Galen Literary Agency, Inc.

frame). Bodies became smaller, but the brain did not follow. Our intellect might result not, as we flatter ourselves, from the benefits of a great mind, but a small body.

Most of what makes primates what they are resides within the skull. Our relatives are pretty smart: but why? Perhaps it is because the fruit trees favored by many are patchily distributed and a mental map is needed to remember where they are. Social life, too, needs gray matter to tell who is who and how to treat the neighbors. Comparative anatomy hints at the past. The bigger the group, the more complex the society. The size of the brain fits that of the community, with a relationship much better than anything to do with what a particular species eats. Society, not shopping, swelled our heads.

Their contents are expensive. A newborn baby uses more than half its energies on what is in its skull. The human brain is two pounds heavier than that of a similar-sized mammal—a hefty pig, say; but we do not use much more energy than pigs. Men have smaller guts than other primates and have, unlike chickens, invested in brains over bellies. The brain's costs come early on, as the organ reaches almost adult size by the time a child goes to school. As Winston Churchill said, "There is no finer investment for any community than putting milk into babies." A giant intellect explains why so much is needed.

Far more genes are active there than in any other tissue, and they can mutate and evolve as much as others. Many mental disorders have a simple genetic basis. Genes also affect personality, with claims of inborn variation in how sociable or misanthropic people might be and in their ability to deal with words or with tools. When it comes to intelligence, schools mold what nature provides . . . but without some foundation in reason, the most expensive education is wasted. Variation in a gene controlling the growth of cells (brain cells included) plays a small part in differences in intelligence. Other such variants will, no doubt, be found. Whatever might emerge, man's behavior

. . . will always be defined more by habit than by instinct; by what he learns rather than what he is.

The Human Pedigree

Behavior apart, man is—of course—not much more than just another primate. No doubt, the laws of hybridism that keep him separate can be blamed to some degree on intellect. There are rumors of test-tube crosses between men and chimps. One such creature made it to the front pages, although on close examination he turned out to be just an unusually rational ape. A cross between the two primates might at least be possible. Humans and chimps share 98.8 percent of their DNA, humans and gorillas rather less. Their close kinship is proved—as in whales and hippos— by their fellow-travelers. All three carry the same set of viral hitchhikers in the same place in their genes, while those of orangs, gibbons and baboons are more and more different. In chromosomes, too, we are similar. The main difference between the great apes and ourselves is that two chimpanzee chromosomes are fused and three more are reshuffled (which means that any hybrid would be sterile).

Our affinity to chimps should be seen in context. It leaves thousands of genes unaccounted for; and, in any case, the diversity of mammals as a whole hides genetic conservatism, with a mere thirty or so large sections of DNA reordered when humans are compared with cats. One cell surface molecule has changed in man alone. It acts as a docking site for diseases such as malaria (to which humans are uniquely susceptible) and as a means of communication between cells—those in the brain included. Perhaps it played a part in man's struggle for mental supremacy and helped keep him distinct from his kin.

Man may differ little in his anatomy from other primates, but his past has a certain interest if only on grounds of familiarity. Darwin was cautious in *The Origin*, but in *The Descent of Man*, published twelve years later, could afford to be direct: "Man is descended from a hairy quadruped, furnished with a tail and pointed ears, probably arboreal in its habits, and an inhabitant of the Old World." To console those insulted by such a notion, he noted that if the human pedigree was no longer "of noble quality," at least it was "of prodigious length."

Now, we have fossils as proof of his claim. The human record is, like all others, incomplete, although it has become much less so in the past half-century. Dozens of Neanderthals have been found. Another almost complete set of bones comes from the famous "Lucy," who lived more than three million years ago. These are exceptions. Primate remains are as liable to upheaval as are those of less noble beings. Hundreds of fossil teeth, and not much more, are found along the Lower Omo River in Ethiopia, because teeth are tough and last when other relics have gone. Most human fossils are small fragments of bone, upon each of which great mountains of theory have been built.

Lucy, for example, was smaller than her fellows. Perhaps, some suggest, males were then much larger than females and life was based on sexual battles. However, it now seems that there were two different-sized species around at the time, so that her supposedly unequal society disappeared as soon as more fossils were found. Other finds have also caused much speculation. On one set of remains was built a theory of man as a killer ape . . . because the skulls were damaged and bones scored as if by knives. In fact, the cracks were the accidents of time, and the cuts are grooves made as bones tumbled down a riverbed.

Why Vertical and Not Horizontal?

Science needs theories, but—with its shortage of facts— human evolution has rather more than it can cope with. Take the art of walking. Lucy and her predecessors could stand erect (although the organ of balance in her inner ear is more like that of apes than humans, so that perhaps they did not do so very well). It is still a delicately poised talent, as anyone who overdoes the Chardonnay finds out. But why did we change from horizontal animals who, now and again, took to the vertical, to an upright beast who must struggle against gravity with a body that evolved at ninety degrees to where it usually finds itself?

In the days when man was seen as striving to reach a manifest destiny, the answer was obvious: our ancestors stood up to free their hands for tasks more important than walking. Of course, the idea is foolish: it is like saying that the brain evolved in order to watch television. Evolution did not plan ahead to the days of *The Jerry Springer Show.* Man stood up, as he did everything else, for pragmatic and short-term reasons. But what could they have been?

Perhaps a vertical animal found it easier to hunt, or to follow herds of grazers in the hope of plunder. Perhaps our ancestor was himself hunted and, by getting on his hind legs, could see danger in good time. There may have been a change of diet toward fruit high on trees or, as the climate dried, a change in habits as food became more scattered and had to be carried back to a distant camp. Being erect is a sexual display, and a male who could keep upright for longer might do better with the females. Those who lie down on a hot day expose more to the sun (which is why sunbathers rarely do so standing up) and, in the tropics at least, pay the price. All these ideas are reasonable but, given that the event happened five million years ago, are hard to test. With so little from the past, anthropology is one of the few sciences in which it is possible to be famous for having an opinion, and until more facts emerge such speculation is bound to remain.

Incomplete as it is, and overinterpreted as it may be, the record of the past is forceful evidence of the reality of human evolution. We descend, with all other mammals, from a rat-sized creature of a hundred and sixty million years ago whose descendants lived modest lives around the feet of the dinosaurs until those giants were wiped out. The first fossil primates are found soon after that event in the warm and wet Africa of sixty-five million years ago. Some six thousand kinds have lived since then (and two hundred or so remain today, from the quarter-pound mouse lemur to the gorilla, a thousand times heavier). Once, the world had many more species of apes than of monkeys, but now just five great apes are left (one of which is us) while monkeys flourish; proof that there was no inevitable progress toward mankind.

About fifteen million years after the emergence of the first primate, the predecessors of apes appear. A molecular clock based on inserted viruses suggests that the lines to chimp, to gorilla and to humans split some five million years ago. A dozen or so hominines—as the branch upon which we belong is called—have lived since then (although how many are real entities is hard to assess). Over that period, there appeared *Ardipithecus* (four and a half million years old and of intermediate form); several kinds of *Australopithecus* (an African primate from around a million years later which was, roughly speaking, a human below the neck but an ape above); and *Homo habilis* (an animal defined as having crossed the cerebral Rubicon of brain size needed to qualify for our own family). It may have been the first tool-user nearly two and a half million years ago. It was followed by *Homo erectus* (a large-brained ape that looked rather like a man). Each seems to have emerged first in Africa and most spread, at one time or another, to Asia and to Europe.

The first members of our own species, *Homo sapiens*, arose about a hundred and fifty thousand years ago as large, thick-skulled, but recognizably human apes. By the time the miseries of the ice ages were over we were, more or less, ourselves, with smaller brains in a thin skull on a slim and elegant body. Why we shrank is not certain. Perhaps, as in the dwarf mammoths on islands, shortage of food did the job, or perhaps a shift to a kinder society cut down the need for a sturdy frame.

Neanderthal DNA

It once seemed natural to arrange the fossils in a sequence that clambered up an evolutionary tree in single file to man. Every find had a name and a place in the hierarchy of the almost-human. Two species, almost by definition, could not live together, as the cultural superior would at once drive out its brutish ancestor.

As more bones turn up, the story becomes less clear. For much of the time, two or more kinds lived together in an uneasy coexistence rather like that of men and chimps today. In East Africa, for instance, two species of large-brained *Homo* lived alongside smaller-brained *Australopithecines* of several types, with perhaps half a dozen forms present at once. Some enigmatic fossils, such as the "Black Skull" found at Lake Turkana, combine primitive and advanced features and suggest that patterns of change were complex indeed.

In spite of a century's claims of the discovery of "missing links," it is quite possible that no bone yet found is on the direct genetic line to ourselves. With so many kinds to choose from, so few remains of each, and such havoc among their relics, none of the fossils may have direct descendants today. The proportion of the people alive even a hundred thousand years ago who contributed to modern pedigrees is small, and the chance that any surviving relic, one among lost billions, belonged to that elite is quite minute.

Neanderthals are the most familiar of fossils. Once dismissed as the remains of diseased Napoleonic soldiers, and then hailed as our immediate predecessors, their true history is one of a dead end on the road of human evolution. With their stocky bodies, they were adapted to cold; and they evolved in Europe or the Middle East, their home until they disappeared thirty thousand years ago. They used stone tools, but in an uninspired way, and stayed apart from their intellectual neighbors.

Although Neanderthals, at first sight so similar to modern humans, were once placed on the last rung before mankind, fossil DNA hints that they may not even be on the same ladder. Their mitochondrial genes are quite distinct from our own. They were not the ancestors of human genes but followed a separate path. For mitochondria, at least, Neanderthals and ourselves split half a million years ago.

Today's molecules also hint that our immediate ancestors have gone forever. The longer the history of any population, the more variation it contains, because there has been more time for mutations to build up. The more abundant it is, the less the chance of an accidental loss of the new variants as they arise. Any large and ancient group of animals hence contains lots of different inherited forms, many of which—given time—become common. A new or sparse population, in contrast, has little diversity and the altered forms are rare.

Genes Provide Clues to the Past

The story of genes is rather like that of surnames. The probable date of the shared ancestor of everyone with a certain surname depends on the size and age of the population. Half a dozen Sidebothams in a tiny village are almost certain to stem from the same recent ancestor. Six carriers of that noble name chosen at random from among the many Sidebothams of London will have to go much further back to find the ur-Sidebotham from whom they descend. A city in which most people have the same name is likely to have expanded from a small village, while one with thousands of surnames traces its origin to a huge and variable ancient populace. Genes, like names, can be used to make guesses about the past.

They show that, compared to other primates, humans are not very diverse and most variants are rare. Chimpanzees are three times more distinct one from the other than are men, with fifty times as much divergence among separate populations. The logic of the genes shows that chimps, now only a couple of hundred thousand strong, were once common and that the human race, abundant as it is today, was scarce. Its average size over most of its history may have been a mere ten to twenty thousand people. Our immediate ancestors were a small band who occupied a few hundred square miles and, more than likely, left no fossils at all. To draw the human family tree reveals an explosion of change, a starburst like that of AIDS viruses, with its center a hundred thousand years ago. It may mark the expansion and spread of modern *Homo sapiens*, but the chances of finding where we came from are small indeed.

Wherever the Garden of Eden may have been, man became a traveler as soon as he escaped it. Although a shortage of bones makes it hard to be certain, there were several journeys out of (and perhaps even back to) Africa.

The great arena of evolution was in Asia, Europe and Africa. The habitable world was not filled until a thousand years ago, with the settlement of New Zealand. Many barriers stood in the way. Some, narrow though they are, proved hard to cross. The New World, most agree, was not reached until about fourteen thousand years ago, across the Bering Land Bridge between what is now Siberia and Alaska. Within a thousand years people had reached southern Chile (where hints have been found of an occupation twice as old). By contrast, humans have been in Australia for fifty thousand years, and crossed the Sunda Strait—recognized by Wallace as a break between the continents—to do so.

The simplest of all barriers is distance. Although that has been defeated by the wheel, it is easy to forget how isolated all of us once were. Numbers shot up after farming began, but even then, most people stayed at home. Although the conventional view of history is of rape and pillage—men, from Attila the Hun onward, forcing their genes on to women—DNA reveals another past. The genes that pass through females, on the mitochondria, are less localized than those of the male chromosome, the Y. Women, it seems, have traveled more than men (perhaps because they move to find a husband in the next village).

Like Red Sea fish in the Mediterranean, men destroy as they move on. Alfred Russel Wallace noted that "we live in a zoologically impoverished world, from which all the hugest, and fiercest, and strangest forms have recently disappeared." The culprit is plain. Humans reached Australia fifty thousand years ago. They came across tortoises as big as a Volkswagen Beetle, carnivorous kangaroos, and flightless birds twice the size of an emu. Within ten thousand years, all were gone. The first Americans were even more efficient. Throwing sticks, atlatls, increased the leverage on a spear and gave it the power of a Magnum rifle. Five hundred years after the hunters reached the Great Plains, mammoths, camels and horses were extinct or almost so and the survivors—wild sheep and bison—were much reduced in size. Their demise is a reminder that, for most species, extinction at the hands of a successor is inevitable.

Travel as he might, man is a lowland animal. Eleven of the world's fifteen largest cities are on the coast. Mountain—and northern—peoples have been pushed around by ice ages as much as have dung beetles. In China, the tribes of the hills are distinct while the masses of the coastal plains are more uniform. Even in Italy, hill villages are, because of their isolation, more different from one another in their genes than are the cities of the plain.

DNA and Human Evolution

DNA reconstructs the history of human migration. The trends across the world reflect bonds of shared descent, as modified by natural selection. With few exceptions, people from northern places are taller and broader than those from the tropics. What counts is the relation between the mass of the body and its surface area. To stay warm it pays to be spherical. Eskimos (and Neanderthals) have thick bodies with short limbs, while most Africans are slimmer with longer arms and legs. Patterns of body shape are consistent to north and south, and those in skin color (little though we understand them) have parallels in the New World and the Old.

On islands, too, people have changed, more by the accidents of migration than by natural selection, as most islands have not been occupied for long. All over the world, isolated by water, by mountains or by bigotry, they have evolved. By chance, certain genes that are rare at home took a trip and at once became relatively common. The inhabitants of several Pacific islands suffer from inherited blindness, and religious isolates such as the Amish of North America also have much genetic disease.

Human evolution can, of course, be studied in the same objective way as that of any animal. Genetics, geology, and geography illuminate our past. Morphology—comparative anatomy—is the most powerful tool of all. Queen Victoria noticed as much on her first visit to London Zoo, in 1842, and was not amused: "The Orang Outang is too wonderful . . . He is frightful and painfully and disagreeably human." She was right. Anti-AIDS medicines are tested on chimps because their bodies are so like ours. The cladistic rules put humans into the same family as chimps and gorillas, as all derive from a common ancestor not shared by anything else. The laws that put birds with crocodiles cannot be broken simply to satisfy our wish to be in a class of our own.

Man is a primate, and in some ways not a very special one. He can do more than any other creature, but has not changed much to do so. The strangest thing about human evolution is how little there has been. Nothing else is so widespread and nobody fills so many gaps in the economy of nature. Many animals carry out tasks almost as wonderful as those achieved by ourselves, but through biology rather than intellect. For them, success at one task means failure at all others. In the past hundred thousand—in the past hundred—years, human lives have been transformed, but bodies have not. We did not evolve, because our machines did it for us. As Darwin put it in *The Descent of Man:* "The highest possible stage in moral culture is when we recognize that we ought to control our thoughts.". . .

Some of Darwin's ideas survive while others have been proved wrong. *The Origin of Species* endures as a work of art as much as of science. Its message remains. Man, the highest of animals, and the most exalted object which we are capable of conceiving, emerged from the war of nature, from famine and death, as much as did all others. Humans, alone, have gone further. As a result, much of what makes us what we are does not need a Darwinian explanation. The birth of Adam, whether real or metaphorical, marked the insertion into an animal body of a post-biological soul that leaves no fossils and needs no genes. To use the past to excuse the present is to embrace [the] fallacy, that society can be explained in terms of the animal world. However, the new insight that biology gives into our history releases us from the narcissism of a creature that is one of a kind. It shows that humans are part of creation, because we evolved.

Chronology

B.C.

ca. 495–435
The life of the Greek philosopher-scientist Empedocles, who proposes an early version of the theory of evolution, including the kernel of the concept of "survival of the fittest."

ca. 99
The birth of the Roman philosopher and poet Lucretius, who repeats and elaborates on the ideas of the Greek evolutionists.

A.D.

1650
Protestant theologian James Ussher declares that God created Earth in 4004 B.C.

1745–1751
French natural philosopher Pierre Maupertuis proposes the most complete and plausible pre-Darwinian theory of evolution.

1794–1796
Charles Darwin's grandfather, Erasmus Darwin, publishes *Zoonomia*, in which he outlines his own version of evolution.

1796
French anatomist George Cuvier proposes that large-scale extinctions of animal species have occurred in the past.

1798

British economist Thomas Malthus publishes his *Essay on Population*, which will later inspire both Charles Darwin and another naturalist, Alfred Russel Wallace, to conceive of the principle of natural selection.

1809

French scientist Jean-Baptiste de Monet de Lamarck introduces his version of evolution; Charles Darwin is born on February 12 in Shrewsbury in western England.

1831

Charles Darwin sails on the vessel HMS *Beagle*, which begins an around-the-world voyage of exploration.

1836

The *Beagle* returns to England.

1837

Darwin publishes a journal of his adventures and discoveries on the *Beagle*'s voyage.

1838

Darwin reads Malthus's population essay and begins formulating the theory of natural selection.

1844

Darwin writes a 230-page outline of his theory and shows it to his friend, the noted botanist Joseph Hooker.

1856

Darwin begins writing his masterwork, *The Origin of Species.*

1859

The Origin of Species is completed and published.

1860

Hooker and noted biologist Thomas H. Huxley successfully

defend Darwin and his theory in a heated debate on evolution held at prestigious Oxford University.

1867

Darwin begins work on *The Descent of Man*, in which he examines human evolution.

1882

Charles Darwin dies at age seventy-three.

1919

A Minnesota minister named William Bell Riley helps to found the World Christian Fundamentals Association, officially launching the American fundamentalist movement.

1921

The fundamentalists try but fail to ban the teaching of evolution in South Carolina schools.

1925

The Butler Act, which prohibits the teaching of evolution in classrooms, passes the Tennessee legislature; John Scopes, a science teacher in Dayton, Tennessee, is tried and convicted of breaking the new law in what becomes known as the "Monkey Trial."

1968

The U.S. Supreme Court rules that laws like the Butler Act are unconstitutional because they violate the First Amendment clause providing for separation of church and state.

1972

Biologists Stephen Jay Gould and Niles Eldredge introduce the idea of "punctuated equilibrium," which states that evolution often occurs in rapid bursts, followed by periods of little change.

1981

The Arkansas legislature passes Act 590, which mandates

that "creation science" must be taught along with evolution in schools; a few months later, a district judge nullifies the law, saying that creation science is not actually science.

1995

The Alabama state board of education decides that all biology textbooks in the state must include a disclaimer saying that evolution is only a theory.

Organizations to Contact

The editors have compiled the following list of organizations concerned with the topics contained in this book. The descriptions are derived from materials provided by the organizations. All have publications or information available for interested readers. The list was compiled on the date of publication of the present volume; the information provided here may change. Be aware that many organizations take several weeks or longer to respond to inquiries, so allow as much time as possible.

American Scientific Affiliation (ASA)
PO Box 668, Ipswich, MA 01938
(978) 356-5656 • fax: (978) 356-4375
e-mail: asa@asa3.org • Web site: www.asa3.org

ASA membership is composed of industrial and academic scientists subscribing to the Christian faith. It seeks to integrate, communicate, and facilitate properly researched science and biblical theology in service to the church and the science community. It seeks to have theology and science interacting in a positive light. Its publications include the *American Scientific Affiliation Newsletter* and *Perspectives on Science and Christian Faith.*

Creation Research Society (CRS)
PO Box 8263, St. Joseph, MO 64508-8263
e-mail: contact@creationresearch.org
Web site: www.creationresearch.org

Persons with at least a master's degree in some branch of science are voting members, and sustaining members are other interested individuals. CRS is for Christians who believe that the facts of science support the revealed account of creation in the Bible. It maintains a laboratory-equipped research center in Arizona (see next entry) and conducts research and disseminates information to the public.

Creation Research Society (CRS)
Van Andel Research Center
6801 North Hwy. 89, Chino Valley, AZ 86323-9186
(928) 636-1153 • fax: (928) 636-8444
e-mail: vacrc@creationresearch.org
Web site: www.creationresearch.org

CRS facilitates and supports the scientific study of the theories of creation and evolution. Its resources include a meteor astronomy observatory, research greenhouse, electronics lab, gas chromatograph, and virtual instrumentation. Its publications include *Creation Research Society Quarterly*.

Genesis Institute (GI)
10220 N. Nevada, Suite 280, Spokane, WA 99218
(509) 467-7913 • fax: (509) 467-0344
e-mail: dave.hutchins@genesisinstitute.org
Web site: www.genesisinstitute.org

GI is made up of individuals seeking to publicize the value of the Gospels in sciences and bring the Bible and science together. It stresses creation evangelism and believes that the universe is less than six thousand years old. It conducts educational and research programs and offers home schooling services.

Institute for Creation Research (ICR)
10946 Woodside Ave., Santee, CA 92071
(619) 448-0900 • fax: (619) 448-3469
Web site: www.icr.org

ICR asserts the inerrancy of scripture through the abundant evidence in science. It conducts research and education. Its publications include *Acts and Facts, Days of Praise*. Its educational activities include summer institutes, educational workshops, graduate school courses, lectures, and seminar programs.

Institute of Human Origins (IHO)
Arizona State University
PO Box 874101, Tempe, AZ 85287-4101
(480) 727-6580 • fax: (480) 727-6570
e-mail: iho@asu.edu • Web site: www.asu.edu/clas/iho

The institute is comprised of scientists, educators, students, volunteers, and other individuals carrying out or supporting research on human evolution. It utilizes the expertise and knowledge of many disciplines to establish when, where, and how the human species originated. It promotes laboratory and field research. It provides a base from which research can be pursued from the planning stages to the dissemination of results. It offers specialized training to scientists and students and maintains a repository and data center of photos, slides, casts, field notes, and comparative collections. It compiles statistics and maintains a speakers' bureau.

National Academy of Sciences (NAS)
500 Fifth St. NW, Washington, DC 20001
(202) 334-2000 • fax: (202) 334-2158
Web site: www4.nationalacademies.org/nas/nashome.nsf

The NAS is a private, honorary organization dedicated to furthering of science and engineering; members are elected in recognition of their distinguished and continuing contributions to either of the two fields. Founded by an act of Congress to serve as official adviser to the federal government on scientific and technical matters. Its publications include *Biographical Memoirs* and the monthly *Proceedings of the National Academy of Sciences*.

National Center for Science Education (NCSE)
420 Fortieth St., Suite 2, Oakland, CA 94609-2509
(510) 601-7203 • fax: (510) 601-7204
e-mail: ncseoffice@ncseweb.org • Web site: www.ncseweb.org

NCSE is affiliated with the American Association for the Advancement of Science. It is made up of scientists, teachers, students, clergy, and interested individuals. NCSE seeks to improve science education, specifically the study of evolutionary science, and opposes the teaching of creationism as part of public school science curricula. It publishes books, pamphlets, and audio and video cassettes on evolution education and education on the nature of scientific inquiry. It also publishes *Reports of the National Center for Science Education*. It reaches markets through direct mail and accepts unsolicited manuscripts on evolution and science education.

National Science Teachers Association (NSTA)
1840 Wilson Blvd., Arlington, VA 22201-3000
(703) 243-7100 • fax: (703) 243-3924
Web site: www.nsta.org

NSTA is made up of teachers seeking to foster excellence in science teaching. It studies students and how they learn, the curriculum of science, the teacher and his/her preparation, the procedures used in classroom and laboratory, the facilities for teaching science, and the evaluation procedures used. Its publications include *Journal of College Science Teaching, Reports on the Teaching of Science at the College Level.* It also publishes curriculum development and professional materials, teaching aids, career booklets, and audiovisual aids.

Reasons to Believe (RTB)
PO Box 5978, Pasadena, CA 91117
(800) 482-7836 • (626) 335-1480 • fax: (626) 852-0178
Web site: www.reasons.org

RTB seeks to explain the theory of creation in a biblically sound and scientifically valid manner, in an effort to remove the doubts of skeptics and strengthen the faith of Christians. It conducts research and educational programs and operates a speakers' bureau. Its publications include the quarterly newsletter *Facts and Faith.*

For Further Research

Books

John F. Ashton, ed., *In Six Days: Why Fifty Scientists Chose to Believe in Creation*. New York: Master, 2001.

Michael J. Behe, *Darwin's Black Box: The Biochemical Challenge to Evolution*. New York: Free, 1996.

John Bowlby, *Charles Darwin: A New Life*. New York: W.W. Norton, 1990.

Peter J. Bowler, *Evolution: The History of an Idea*. Berkeley and Los Angeles: University of California Press, 1989.

John L. Brooks, *Just Before the Origin: Alfred Russel Wallace's Theory of Evolution*. New York: Columbia University Press, 1984.

Charles Darwin, *The Origin of Species*. New York: New American Library, 1958.

Erasmus Darwin, *Zoonomia: Or the Laws of Organic Life*. 2 vols. New York: AMS, 1974.

Francis Darwin, ed., *Charles Darwin: His Life Told in an Autobiographical Chapter and in a Selected Series of His Published Letters*. New York: Appleton, 1892.

Richard Dawkins, *The Blind Watchmaker: Why the Evidence of Evolution Reveals a Universe Without Design*. New York: W.W. Norton, 1987.

Daniel C. Dennett, *Darwin's Dangerous Idea: Evolution and the Meanings of Life*. New York: Simon & Schuster, 1995.

Michael Denton, *Evolution: A Theory in Crisis*. Bethesda, MD: Adler and Adler, 1985.

Adrian Desmond and James Moore, *Darwin.* New York: Warner, 1992.

Duane T. Gish, *Evolution: The Fossils Say No!* San Diego: Creation-Life, 1978.

Bentley Glass et al., eds., *Forerunners of Darwin: 1745–1859.* Baltimore: Johns Hopkins University Press, 1968.

William Irvine, *Apes, Angels, & Victorians: The Story of Darwin, Huxley, and Evolution.* New York: McGraw-Hill, 1955.

Phillip E. Johnson, *Darwin on Trial.* Washington, DC: Regnery Gateway, 1991.

L.J. Jordanova, *Lamarck.* New York: Oxford University Press, 1984.

Walter Karp, *Charles Darwin and the Origin of Species.* New York: American Heritage, 1968.

Desmond King-Hele, *Erasmus Darwin.* New York: Scribner's, 1963.

Edward J. Larson, *Summer of the Gods: The Scopes Trial and America's Continuing Debate over Science and Religion.* Cambridge, MA: Harvard University Press, 1998.

Lucretius, *On Nature*, published as *Lucretius on the Nature of the Universe.* Trans. Ronald Latham. Baltimore: Penguin, 1962.

Tom McGowen, *The Great Monkey Trial: Science vs. Fundamentalism in America.* New York: Franklin Watts, 1990.

Don Nardo, *Charles Darwin.* New York: Chelsea House, 1993.

———, *The Scopes Trial.* San Diego: Lucent, 1997.

National Academy of Sciences, *Science and Creationism.* Washington, DC: National Academy, 1999.

Chet Raymo, *Skeptics and True Believers: The Exhilarating*

Connection Between Science and Religion. New York: Walker, 1998.

Michael Ruse, *The Evolution Wars: A Guide to the Debates.* Santa Barbara, CA: ABC-CLIO, 2000.

Alfred Russel Wallace, *Darwinism: An Exposition of the Theory of Natural Selection.* New York: AMS, 1975.

Robert Wesson, *Beyond Natural Selection.* Cambridge, MA: MIT Press, 1991.

Periodicals

Larry Arnhart, "Darwin's Science of Morality," *American Outlook,* November/December 2000.

Jerry Bergman, "A Brief History of the Modern Creationist Movement," *Contra Mundum,* Spring 1993.

Mona Charen, "Scopes Trial Replayed in Kansas," *Conservative Chronicle,* August 25, 1999.

Frederick C. Crews, "Saving Us from Darwin," *New York Review of Books,* October 4, 2001.

Thomas J. Geelan, "When Creationism Goes to School," *Free Inquiry,* Spring 2000.

Wendy Kramer, "Are We Evolved Yet?" *Free Inquiry,* Fall 2000.

Michael Ruse, "Answering the Creationists," *Free Inquiry,* Spring 1998.

Adam Sedgwick, "Objections to Mr. Darwin's Theory of the Origin of Species," *Spectator,* April 7, 1860.

John M. Swomely, "On Creationism and Evolution," *Human Quest,* March/April 2000.

Internet Sources

Steve Case, "Why Evolution Must Not Be Ignored," 1999.

www.washingtonpost.com/wp-srv/national/zforum/
99/nat082799.htm.

Lenny Flank, "Creation 'Science': A Legal History," 2003.
www.geocities.com/CapeCanaveral/Hangar/2437/legal.
htm.

———, "Does Science Discriminate Against Creationists?"
1995. www.geocities.com/CapeCanaveral/Hangar/2437/
discrimination.htm.

Institute for Creation Research, "ICR Tenets of Creationism," 2003. www.icr.org/abouticr/tenets.htm.

Mark Isaak, "Five Major Misconceptions About Evolution,"
2003. www.talkorigins.org/faq/faq-misconceptions.html.

Donald Kaul, "Creationists Make God in Their Image," 1999.
www.holysmoke.org/cre-nons.htm.

Douglas Linder, "Famous Trials in History: Tennessee vs.
John Scopes: The 'Monkey Trial,' 1925," 2002. www.law.
umkc.edu/faculty/projects/ftrials/scopes/scopes.htm.

Lyndsey McCabe, "The Scopes 'Monkey Trial'—July 10,
1925–July 25, 1925," 1996. http://xroads.virginia.edu/
~ug97/inherit/1925home.html.

Roger W. Sanders, "Rebuttal to Anthony Browne, 'Scruffy Little Weed Shows Darwin Was Right as Evolution Moves
On,'" 2003. www.icr.org/headlines/weedevolution.html.

Kerwin Thiessen, "Should Creationism Be Taught in Public
Schools? Yes!" 2003. www.directionjournal.org/article/
?449.

Historian Don Nardo has published many books for young adults about the history of science and new scientific discoveries, including *Greek and Roman Science, The Extinction of the Dinosaurs, Black Holes*, and *Extraterrestrial Life*. He has also written extensively about Charles Darwin, the theory of evolution, and the Scopes trial. Mr. Nardo lives with his wife, Christine, in Massachusetts.